ARTIFICIAL INTELLIGENCE METHODS AND TOOLS FOR SYSTEMS BIOLOGY

Computational Biology

VOLUME 5

ARTIFICIAL INTELLIGENCE METHODS AND TOOLS FOR SYSTEMS BIOLOGY

Edited by

WERNER DUBITZKY

and

FRANCISCO AZUAJE

A C.I.P. Catalogue record for this book is available from the Library of Congress.

ISBN 1-4020-2959-4 (PB)
ISBN 1-4020-2859-8 (HB)
ISBN 1-4020-2865-2 (e-book)

Published by Springer,
P.O. Box 17, 3300 AA Dordrecht, The Netherlands.

Sold and distributed in North, Central and South America
by Springer,
101 Philip Drive, Norwell, MA 02061, U.S.A.

In all other countries, sold and distributed
by Springer,
P.O. Box 322, 3300 AH Dordrecht, The Netherlands.

Printed on acid-free paper

Printed in the Netherlands.

'Entities of an essentially new sort are entering the sphere of scientific thought. Classical science in its diverse disciplines, be it chemistry, biology, psychology or the social sciences, tried to isolate the elements of the observed universe —chemical compounds and enzymes, cells, elementary sensations, freely competing individuals, what not—expecting that, by putting them together again, conceptually or experimentally, the whole or system—cell, mind, society—would result and be intelligible. Now we have learned that for an understanding not only the elements but their interrelations as well are required ...'

Ludwig von Bertalanffy, 1901–1972.

'The whole of the developments and operations of analysis are now capable of being executed by machinery ... As soon as an Analytical Engine exists, it will necessarily guide the future course of science.'

Charles Babbage, 1791–1871.

Contents

Preface

Due to political and socio-economic changes and radical technological advances (e.g. automation, nanotechnology, robotics, computerization) the last decade has revolutionized science and technology and has radically changed the way we apply, manage, and evolve information and knowledge. This is particularly true for the new interdisciplinary scientific field called *systems biology*. Investigating life from its molecular basis all the way up to consciousness, populations, and ecosystems, systems biology covers a staggering range of scientific concepts, methods, and natural phenomena. In addition to systems science, mathematics, and computer science, key sub-disciplines of systems biology include biological and medical physics, chemistry and biochemistry, biology (e.g. evolutionary, developmental, and cell biology), physiology, neuroscience, psychology, behavioral sciences, and medicine. Life science in general and systems biology in particular are currently witnessing a knowledge and information revolution and proliferation on an unprecedented scale. We see two main reasons for this development. First, the ever-increasing detail and sophistication in which natural systems are being probed, mapped, and modeled. And second, a shift from a traditionally 'reductionistic' to a more holistic, integrative, system-based approach to understanding the dynamics and organizational principles of living systems.

The increasingly detailed investigations in systems biology have lead to astronomical mountains of data and information. At a conceptual level, we distinguish two kinds of data repositories: databases and information bases.

- Databases (or data sets) contain 'raw' observational or experimental data, arising from *in vitro*, *in vivo*, and increasingly from *in silico* (i.e. computer simulation) experiments. Examples include gene expression repositories, brain scan data, and toxicity screen data. While databases from systems biology experiments are often local and proprietary, more and more databases are now becoming publicly available.
- Unlike databases, information bases contain *summarized*, *consolidated*, and *derived* data, and are typically organized to serve a particular function. Such repositories or systems organize and manage insights obtained from analyzing and

interpreting experimental data. Most of these repositories are publicly available, for example, gene and protein information bases such as GenBank, Swiss-Prot, and ontologies such as the Gene Ontology.

The push towards a systems approach has profound implications for the way the life sciences will develop in the future and in terms of the role computing science (including artificial intelligence) will play in this development. The new systems thinking in biology has triggered the need for

- constructing more 'complete' *computational models* that are able to capture crucial *knowledge* about and vital *systemic properties* and *mechanisms* of biological entities and processes. Besides reflecting crucial structural characteristics of the underlying biological objects, such models are endowed with inferential capabilities. Systems of this kind operate either as *knowledge-based* or *decision support systems* (capable of intelligently answering non-trivial questions about a well-defined small area of systems biology) or as simulation systems. The former class of systems is often geared towards processing 'symbols' (qualitative, i.e., words) using a logics-based scheme for drawing inferences. This allows them to intelligently answer non-trivial questions about a well-defined small area of systems biology. The latter uses 'sub-symbolic' (quantitative, i.e., numbers) representations and inference mechanisms, facilitating the *simulation* of the systemic properties and dynamics of the biological system under study. Computational models are considered essential in the quest to a deep understanding of life phenomena,
- developing integrative information infrastructures, methodologies, tools, and systems that stretch across the entire data-information-knowledge spectrum and transcend the man-machine dichotomy, and
- bridging the cultural, conceptual, and technological gap among systems biology disciplines.

An important dimension of the information and knowledge proliferation issue in systems biology is geography. Systems biology organizations and information repositories and systems are dispersed around the globe. Many future solutions tackling the information and knowledge needs in systems biology will have to address this dimension. *Grid computing* technologies are hailed as the 'knowledge layer' of the global computing network. As such they are destined to play a crucial role in this endeavor. Methodologies from the field of *distributed artificial intelligence* (agents) and *artificial life* (self-organizing and emergent systems) are important contenders in making contributions to this development.

The opinions on how to define the field of artificial intelligence vary widely, and sometimes wildly. For this title we adopt a fairly broad and practical view. In this view

artificial intelligence *embraces concepts, methodologies, and techniques that form part of a computer system or program that exhibits characteristics akin to intelligent behavior*

If, for example, a program plays chess on a sufficiently high level, or successfully and consistently derives useful and non-trivial patterns from data, or if it predicts the secondary structure of proteins with a satisfactory degree of accuracy, then we would consider it 'artificial intelligence', regardless of the underlying technology. Some key areas of artificial intelligence include:

- Computational *knowledge (bases)* and *reasoning* (under uncertainty). This class of methodologies is used to explore computational ways of representing (potentially incomplete, inconsistent) knowledge about the world and algorithms that reason logically using this knowledge base.
- *(Machine) Learning*. This notion embraces a range of methodologies that aim to automate the generation of knowledge and reasoning structures needed for automated reasoning.
- *Problem solving* and *planning*. These areas are concerned with tasks that require to think ahead several steps, to construct sequences of actions towards a set goal, and to decide which course of action to choose. Methods of this kind employ knowledge and reasoning and learning approaches.
- Agents. Agent techniques model how an artificial intelligent agent perceives and communicates with its environment by vision, touch, hearing, or understanding language, and how agents turn their plans into actions by affecting the environment, for example, using robot motion or language utterances or by providing decision support.

We are convinced that artificial intelligence has a great deal to offer to ongoing and future research and development in systems biology. First, the range of methodologies and tools developed by artificial intelligence researchers may provide solutions to systems biology problems where conventional mathematical and statistical methods prove ineffective or inefficient. Second, artificial intelligence is arguably a key technology for addressing the ever-increasing data and knowledge proliferation problem in systems biology and related life science fields. We envisage that artificial intelligence will make significant contributions to this end (a) by helping to capture, share, evolve, and use knowledge about vast amounts of life phenomena (potential techniques include artificial life, machine learning, knowledge management, multi-agent systems), (b) by providing a methodological and technological platform for integrating and thus optimizing capabilities of humans and computers (e.g., agents, human-computer interaction, natural language processing, vision and pattern recognition, computational creativity), (c) by offering a conceptual and methodological 'pathway' from data to knowledge (e.g., data mining, knowledge-based systems, computational theories), and (d) by playing an important role in the industrial design and development of new compounds, processes, and applications.

We believe that one significant barrier to the widespread use of artificial intelligence methods and tools in systems biology is a lack of knowledge about what kinds of methodologies and tools exist, how such techniques are used, what their merits and limitations are, and what obstacles are involved in deploying them. An important goal of this volume is to address these issues by providing what is simultaneously a *design blueprint*, *user guide*, *research agenda*, and *communication platform* for current and future developments in the field. We intentionally adopt a relatively broad definition of systems biology, embracing a wide range of topics from biological physics to consciousness research. In this definition the goal of

> ***systems biology*** *is to reconcile the fast-growing volumes of data about biomolecules, cells, tissues, organisms, populations, and ecosystems into coherent and systemic views of organization*

However, in this volume we do place an emphasis on the molecular dimension of life phenomena and in one chapter on anatomical and functional modeling the brain. From an information processing perspective, these areas of research share important characteristics, including (a) the desire to relate *structure* (anatomy) to *function* (physiology), (b) the importance of understanding both *phylogeny* (collective evolution) and *ontogeny* (individual development) of the phenomenon under study, and (c) the need to understand holistically the mechanisms of organization and dynamics of complex biological systems at different levels (i.e., intra- and inter-cellular, and different '-omic' levels such as genomic, transcriptomic, proteomic, metabolomic, and so on).

As **design blueprint**, the book is intended for artificial intelligence researchers, life scientists, statisticians, computer experts, technology developers, managers, and other professionals who will be tasked with developing, deploying, and using information technology in the context of life science research and development.

As a **user guide**, this volume seeks to address the requirement of scientists and researchers to gain an overview and a basic understanding of key artificial intelligence and related methodologies and tools used in life sciences research. For these users, we seek to explain the key concepts and assumptions of the various techniques, their conceptual and computational merits and limitations, and, where possible, give guidelines for choosing the methods and tools most appropriate to the task at hand. Our emphasis is not on a complete and intricate mathematical treatment of the presented methodologies. Instead, we aim at providing the users with a clear understanding and practical know-how of the relevant methods in the context of concrete life science problems.

As a **research agenda**, the book is intended for computer and life science students, teachers, researchers, and managers who want to understand the state of the art of the presented methodologies and the areas in which gaps in our knowledge demand further research and development. To achieve this, we have attempted to cover a representative range of life science and systems biology areas and artificial intelligence techniques, and we have endeavored to put together a balanced mix of papers with review, method, and application character. Our aim was to maintain readability and accessibility of a textbook throughout the chapters, rather than compiling a mere reference manual. Therefore, considerable effort was made to ensure that the presented material is supplemented by rich literature cross-references to relevant work.

The book is also intended as a **communication platform** seeking to bridge the cultural, conceptual, and technological gap among key systems biology disciplines (i.e., biology, mathematics, and information technology). To support this goal, we have asked the contributors to adopt an approach that appeals to audiences from different backgrounds.

Clearly, we cannot expect to do complete justice to all four objectives in a single book. However, we do believe that we have succeeded in taking constructive steps toward each goal. In doing so, we hope to advance the understanding of artificial intelligence methodologies and tools relevant to life sciences and systems biology research. In addition, we hope to outline future challenges and opportunities of artificial intelligence and important complementary technologies such as Grid computing, information visualization, high-performance computing, to name but a few.

The design and subsequent deployment and application of artificial intelligence methods and tools to specific systems biology problems rests on the scientific expertise of the researchers involved and their knowledge about the underlying concepts and systems. The available machinery of artificial intelligence methods ranges from machine learning and knowledge-based systems (e.g., decision trees, expert systems), to highly mathematical models (e.g., Bayesian statistics and hidden Markov models), and to sophisticated approaches exploiting principles of self-organization and complexity emergence, including multi-agent systems and artificial life techniques. Likewise, the phenomena investigated and methods used by biologists cover a staggeringly huge number of complex concepts. Hence, the preparation of this book must draw upon the experts from many diverse subfields in artificial intelligence, computer science, and life sciences. In developing this volume, we have assembled a distinguished set of authors, each recognized as an authority in one or more of these fields. We have asked these authors to present a selected set of current systems biology research questions, issues, and areas and to describe how artificial intelligence techniques can help to address them. By emphasizing a highly user-oriented and practical approach, the authors have attempted to provide valuable and useful knowledge to researchers and developers dealing with such problems. To support the research agenda of this book, we have also asked the authors to identify where future developments are likely to take place and to provide a rich set of pointers to theoretical and practical works underpinning the presented material. Finally, we invited the authors to design their contributions in such a way that it will encourage and foster dialog across disciplines and (perceived) boundaries. The result, we hope, is a book that will be valuable for a long time, as summary of where we are, as a practical user guide for making informed choices on systems biology projects, and as roadmap for where we need to go in order to improve and further develop future information technology in the life sciences.

This book contains eleven technical chapters, dealing with specific areas of systems biology (see table below). In terms of the underlying life science field, they may be roughly grouped into three broad categories, namely *biochemistry* (first two chapters), (*molecular*) *biology* (Chapters 3 to 10), and *neuroscience* (Chapter 11). Organizing the chapters in this way roughly mirrors the 'systems stack' of life—biochemistry, physiology, and psychology. Within each category, the chapters are presented in an order that reflects as much as possible the degree of complexity and organization of the investigated phenomenon. The table also illustrates the different artificial and information technology techniques discussed in the various chapters.

Table 1. Organization of chapters and systems biology and information technology topics and techniques.

Chapter Number	Chapter Title	Systems Biology	Information Technology
1	Lazy Learning for Predictive Toxicology based on a Chemical Ontology	toxicology	machine learning
2	QSAR Modeling of Genotoxicity on Non-Congeneric Sets of Organic Compounds	genotoxicity	machine learning, QSAR, QSPR
3	Characterizing Gene Expression Sequences using a Hidden Markov Model	gene expression	HMM
4	Analysis Of Large-Scale mRNA Expression Data Sets By Genetic Algorithms	gene expression	genetic algorithms, machine learning
5	A Data-Driven, Flexible Machine Learning Strategy for the Classification of Biomedical Data	proteomics, mass spectroscopy	machine learning, classification
6	Cooperative metaheuristics for exploring proteomic data	proteomics	machine learning, evolutionary algorithms
7	Integrating gene expression data, protein interaction data, and ontology-based literature search	proteomics, gene expression	info integration and retrieval, text mining
8	Ontologies in Bioinformatics and Systems Biology	bioinformatics	ontologies, knowledge-based techniques
9	Natural Language Processing in Systems Biology	annotation, database curation	NLP, text mining, ontologies
10	Systems Level Modeling of Gene Regulatory Networks	gene regulation and expression	systems modeling, Bayesian networks
11	Computational Neuroscience for Cognitive Brain Functions	neuroscience	systems modeling

Chapter 1 discusses the problem of predicting toxic properties of chemical compounds. It reviews several machine learning approaches, which traditionally represent compounds and properties as vectors of attribute-value pairs. The authors propose a new methodology to represent this type of data, which may exploit *structure-activity relationships* (the dependence of the biological effects of a chemical upon its molecular structure produces a structure-activity relationship) in a more meaningful fashion. Such an information representation model is based on the application of knowledge derived from a *chemical ontology*. This model and similarity-based machine learning methods are assessed. The predictive performance obtained is comparable to that shown by other methods based on structure-activity relationships.

In contrast heavily-researched congeneric toxic chemicals (i.e., chemicals belonging to well-defined families of molecular structures), **Chapter 2** investigates the toxic properties of *non-congeneric* substances. This class of compounds constitutes a much harder problem. It reviews advances in the prediction of *genotoxicity* (refers to the damage a toxin can inflict on DNA molecules) of non-congeneric, organic compounds. Comparisons between important predictive methods, such as neural-network-based methods, are implemented and evaluated. This study suggests that neural networks can provide the basis for more powerful structure-activity prediction models than traditional linear techniques.

Chapter 3 moves up to the *transcriptomic* level of the 'systems stack' of life and addresses the problem of describing and classifying *temporal gene expression* data (mRNA abundance levels measured at different time points). The authors apply *hidden Markov models* to detect important temporal patterns in the data. This approach is based on *Markov chains* that model sequences of events where the probability of an event occurring depends on the occurrence of the preceding event. The authors present a procedure for the identification of linguistic identifiers, which may support annotation of gene expression data. They discuss current limitations and requirements for understanding complex, non-obvious temporal relationships.

The selection of the most relevant predictive characteristics or features (e.g., genes in gene expression analyses) for building accurate and efficient classifiers is a fundamental problem in genome-wide studies. **Chapter 4** reviews powerful optimization techniques based on *genetic algorithms* to perform feature selection in gene expression studies. Genetic algorithms implement optimization and search inspired by key principles of evolution (inheritance, variation, and selection based on fitness). The authors discuss design principles, which are important to understand when and how to apply a particular option. Different techniques for information representation, search and classification are compared. Current limitations, strengths and recommendations for future research are offered.

Chapter 5 addresses the problem of improving classification tasks for biomedical spectral data, such as those acquired from mass spectroscopy and magnetic resonance experiments. The authors discuss two important factors: The *curse of dimensionality* and *curse data sparsity*. This double curse is looming large over modern system-wide high-throughput studies in systems biology. It refers to the undesired situation where the number of individual measurements taken per individual sample (or observation) is extremely high while at the same the number available samples is relatively small. Gene expression experiments, for example, are frequently visited upon by this twin evil. They describe a classification framework known as *statistical classification strategy* to address problems of this nature. Diverse pre-processing and classification procedures are introduced and compared. The authors argue that this framework is able provide robust and meaningful classification solutions for complex classification problems (e.g., classification of mass-spec data).

Metaheuristic methods represent an intricate yet sophisticated strategy to address combinatorial optimization tasks, such as *multiple sequence alignment*. Multiple sequence alignment is concerned with establishing a degree of similarity, homology, or other degree of relatedness between two or more sequential biomolecules (usually nucleotide or amino acid sequences) based on a an 'aggregated match' score of the compared sequences. The metaheuristic represents this problem as search of an extremely large search space and uses several processing entities to explore this space. These entities exchange information among them to approximate optimal solutions. *Co-operative metaheuristic processes* may support a more effective and efficient optimization based, for example, on evolutionary or genetic computation algorithms. **Chapter 6** discusses advances in co-operative metaheuristic techniques, which are tested on protein identification (given a mixture containing N different types of proteins, protein identification refers to the task of determining which types these are; this task becomes highly complex if some of the proteins in the mixture are novel) and multiple sequence alignment tasks. Advantages, challenges and design factors are discussed to promote the application of this approach.

A fundamental task in systems biology is the integration of multiple data sources to support hypothesis generation and validation. Artificial intelligence together with advances from networking and Grid computing technologies could facilitate the development of information infrastructures required to tackle these challenges. **Chapter 7** presents a platform for linking gene expression, interaction (gene-gene, gene-protein, protein-protein), and scientific literature data to achieve a systemic visualization of interaction networks. The authors discuss different data analysis tools including clustering and ontology-driven literature search. The application and usefulness of this framework is illustrated by an analysis of energy-related genes and protein complexes, that is, molecules involved in the transformation, exchange, and use of energy in cellular processes.

Data integration, literature search, natural language processing and annotation tasks (characterization of genes, proteins and other biological entities and processes) require a profound understanding of how biology knowledge may be formally represented and harmonized. *Biological ontologies* have become a crucial tool for representing and sharing knowledge in systems biology. A biological ontology refers to an explicit formal specification of how to represent the objects, concepts and other entities that are assumed to exist in (systems) biology or its sub-disciplines and the relationships that hold among these entities. **Chapter 8** introduces the problem of constructing and exploiting ontology-driven knowledge in systems biology. The author provides and overview and conceptual background and discuss design tools and key applications in bioinformatics, for example, in complex database querying systems.

Chapter 9 reviews applications of *natural language processing* (NLP) for supporting various tasks relevant to systems biology. Natural language processing is concerned with the use of computers to process written and spoken language for some practical and useful purpose. The chapter discusses the connection between NLP techniques and complex biological problems. It illustrates important NLP-based systems and applications, including the curation of biological databases and data mining for functional genomics. Database curation is concerned with the need to maintain the correctness, consistency and currency of biomedical databases in the light of new facts and emerging knowledge. The computational requirements, limitations and design issues for achieving advanced NLP systems in bioinformatics are also discussed.

Chapter 10 overviews the application of neural networks and statistical learning approaches to aid the modeling of *genetic networks*. Genetic networks govern which genes are expressed (active) in a cell at any given time, how much product (RNA or protein) is made from each one, and the cell's responses to diverse environmental cues and intracellular signals. Fundamental concepts on inferring genetic networks from gene expression data are introduced. The chapter places emphasis on data-driven methods, such as clustering-based and graphical modeling methods, for the study of genetic networks. The authors also discuss the important problem known as *generative inverse modeling*, in which gene expression data can be derived (generated) by simulation from computational models (here: *Bayesian networks*) representing genetic networks.

Finally, **Chapter 11** describes and evaluates a computational, integrative framework for studying brain function. The approach described in this chapter integrates information originating from different levels of higher brain function. This model may be used to implement, for example, simulations of neuronal responses or the effect of pharmacological agents (e.g., therapeutic drugs). The presented analyses show how an important cognitive factor known as attentional bias can affect other important processes such as selective working memory. *Attentional bias* refers to the ability of the brain to cope with the massive amount of sensory input via selective attention causing certain inputs to be processed in a preferential fashion.

The book is designed to be used by the practicing professional tasked with the organization and analysis of life science and systems biology data, and the modeling of life phenomena and systems. It is also intended to serve as a text for a senior undergraduate- or graduate-level course in cheminformatics, bioinformatics, neuroinformatics, complex systems studies, systems biology, or topics on artificial intelligence. In a quarter-length course, one lecture can be spent on each chapter, and a project may be assigned based on one of the topics or techniques discussed in a chapter. In a semester-length course, some topics can be covered in greater depth, covering more of the formal background of the discussed methods. Each chapter includes recommendations for further reading.

Acknowledgments

First and foremost we should like to acknowledge the contributors for their support and patients. Also, we extend our thanks to Jesús A. López, who offered excellent advice on intricate matters regarding LaTeX coding and debugging.

Coleraine and Jordanstown, *Werner Dubitzky*
May 2004 *Francisco Azuaje*

Lazy Learning for Predictive Toxicology based on a Chemical Ontology

Eva Armengol and Enric Plaza

Artificial Intelligence Research Institute (IIIA-CSIC)
Campus UAB, 08193 Bellaterra, Catalonia (Spain).
E-mail: {eva, enric}@iiia.csic.es

Summary. Predictive toxicology is concerned with the task of building models capable of determining, with a certain degree of accuracy, the toxicity of chemical compounds. We discuss several machine learning methods that have been applied to build predictive toxicology models. In particular, we present two lazy learning lazy learning techniques applied to the task of predictive toxicology. While most ML techniques use structure relationship models to represent chemical compounds, we introduce a new approach based on the chemical nomenclature to represent chemical compounds. In our experiments we show that both models, SAR and ontology-based, have comparable results for the predictive toxicology task.

1 Introduction

Thousands of new chemicals are introduced every year in the market for their use in products such as drugs, foods, pesticides, cosmetics, etc. Although these new chemicals are widely analyzed before commercialization, the effects of many of them on human health are not totally known. In 1973 the European Commission started a long term program consisting on the design and development of toxicology and ecotoxicology chemical databases. The main idea of this program was to establish lists of chemicals and methods for testing their risks to the people and the environment. Similarly, in 1978 the American Department of Health and Human Services established the National Toxicology Program (NTP) with the aim of coordinating toxicological testing programs and developing standard methods to detect potentially carcinogenic compounds (see more information in http://www.ntp-server.niehs.nih.gov).

When a chemical compound is suspected to be toxic, it is included in the NTP list in order to perform standardized experiments to determine its degree of toxicity. Basically, there are two kinds of experiments: *in vitro* and *in vivo*. *In vitro* experiments are carried out on *salmonella* and the outcome are quantitative results of several physical-chemical parameters. *In vivo* experiments are performed on rodents (rats and mice), and there are, in turn, two kind of experiments: short-term (90 days) and long-term (2 years). Usually, short-term experiments are performed as a means to

W. Dubitzky and F. Azuaje (eds.), Artificial Intelligence Methods and Tools for Systems Biology, 1–18.

obtain a first clue of the toxicity of a compound. It should be emphasized that to determine the toxicity of chemical compounds on rodents is an expensive process that, in addition, offers results that are not conclusive concerning the toxicity in humans.

The use of computational methods applied to the toxicology field could contribute to reduce the cost of experimental procedures. In particular, artificial intelligence techniques such as knowledge discovery and machine learning (ML) can be used for building models of compound toxicity (see [18] for an interesting survey). These models reflect rules about the *structure-activity relationships* (SAR) of chemical compounds. Such rules are used to predict the toxicity of a chemical compound on the basis of the compound's chemical structure and other known physical-chemical properties. The construction of this model is called *predictive toxicology.*

The Predictive Toxicology Challenge (PTC) was a competition held in 1990 with the goal of determining the toxicity of 44 chemical compounds based on both experiments in the lab and the predictive toxicology methods. The results of this challenge [4, 10] showed that the best methods are those taking into account the results of the short-term tests. A second challenge was announced in 1994. This challenge was mainly focused on using ML techniques and results can be found in [30]. The last challenge held in 2001 [19] was also focused on ML techniques and most of them used SAR descriptors. In this challenge most of authors proposed a relational representation of the compounds and used inductive techniques for solving the task.

Currently there still are two open questions in predictive toxicology: 1) the representation of the chemical compounds, and 2) which are the characteristics of a chemical compound that allows its (manual or automatic) classification as a potentially toxic.

In this chapter we describe several approaches to both questions: we propose a representation of the chemical compounds based on the IUPAC (*International Union of Pure and Applied Chemistry*) chemical nomenclature and a *lazy learning technique* for solving the classification task.

2 Representation of chemical compounds

One of the most important issues for developing computational models is the representation of *domain objects*, in our case chemical compounds. In the toxicology domain, there are several key features of the molecule to be taken into account for predicting toxicity. First, there are some concerning to the basic elements of the molecule, such as number of atoms, bonds between atoms, positions, electrical charges, etc. Second, there are physical-chemical properties of the molecule such as lipophilic properties, density, boiling point, melting point, etc. Finally, there often exists prior information about the toxicity of a molecule, which was obtained from studies on other species using different experimental methods.

In the literature, there are two approaches to represent chemical compounds: 1) those representing a compound as a vector of molecular properties (*propositional representation*), and 2) those explicitly representing the molecular structure

```
atom(tr339,1,o,-1).   atom(tr339,2,n,1).    atom(tr339,3,o,0).     atom(tr339,4,c,0).
atom(tr339,5,c,0).    atom(tr339,6,c,0).    atom(tr339,7,c,0).     atom(tr339,8,o,0).
atom(tr339,9,c,0).    atom(tr339,10,n,0).   atom(tr339,11,c,0).    atom(tr339,12,h,0).
atom(tr339,13,h,0).   atom(tr339,14,h,0).   atom(tr339,15,h,0).    atom(tr339,16,h,0).
atom(tr339,17,h,0).   bond(tr339,1,2,1).    bond(tr339,2,3,2).     bond(tr339,2,4,1).
bond(tr339,4,5,1).    bond(tr339,5,6,2).    bond(tr339,5,12,1).    bond(tr339,6,7,1).
bond(tr339,6,13,1).   bond(tr339,7,8,1).    bond(tr339,7,9,2).     bond(tr339,8,14,1).
bond(tr339,9,10,1).   bond(tr339,9,11,1).   bond(tr339,10,15,1).   bond(tr339,10,16,1).
bond(tr339,11,17,1).
atomcoord(tr339,1,3.0918,-0.8584,0.0066).     atomcoord(tr339,2,2.3373,0.0978,0.006).
atomcoord(tr339,3,2.7882,1.2292,0.0072).      atomcoord(tr339,4,0.8727,-0.1152,-0.0023).
atomcoord(tr339,5,0.3628,-1.4003,-0.0094).    atomcoord(tr339,6,-1.0047,-1.6055,-0.0172).
atomcoord(tr339,7,-1.868,-0.5224,-0.0174).    atomcoord(tr339,8,-3.2132,-0.7228,-0.0246).
atomcoord(tr339,9,-1.355,0.7729,-0.0098).     atomcoord(tr339,10,-2.2226,1.8712,-0.0096).
atomcoord(tr339,11,0.018,0.971,0.0028).       atomcoord(tr339,12,1.0343,-2.2462,-0.0092).
atomcoord(tr339,13,-1.3998,-2.6107,-0.0234).  atomcoord(tr339,14,-3.4941,-0.7673,0.8996).
atomcoord(tr339,15,-3.1824,1.7311,-0.0147).   atomcoord(tr339,16,-1.864,2.7725,-0.0043).
atomcoord(tr339,17,0.419,1.9738,0.0087).
```

Fig. 1. Representation of the chemical compound TR-339 using Horn clauses.

of a compound (*relational representation*). In the follow sections we briefly explain these representations (details can be found at `http://www.informatik.uni-freiburg.de/~ml/ptc/` and then we will introduce our own representation based on the chemical ontology used by the experts.

SAR and Qualitative SAR (QSAR) use equation sets that allow the prediction of some properties of the molecules before the experimentation in the laboratory. In analytical chemistry, these equations are widely used to predict spectroscopic, chromatographic and some other properties of chemical compounds. There is a number of commercial tools allowing the generation of these descriptors: CODESSA [22], TSAR (Oxford molecular products, `http://www.accelrys.com/chem/`), DRAGON `http://www.disat.inimib.it/chm/Dragon.htm`), etc. These tools represent a chemical compound as a set of attribute value pairs. This kind of representation is called *propositional* in ML. For instance, the description of a car using propositional description is the following: {(size, medium), (builder, BMW), (model, 250), (color, white)}.

In addition to the knowledge about a particular compound, it is also useful to handle general chemical knowledge, what is called *background knowledge* in ML. Automatic methods that use background knowledge often consider compounds as a structure composed of substructures. This kind of representation is called *relational* because an object is represented by the relationships between their component elements. For instance, a car can be described composed of subparts like the chassis and the engine. In turn, each one of these parts can be described by their own subcomponents.

A form of relational representation is *logic programming*, that represents the relations among elements by a set of predicates. Thus, a set of predicates can be used to establish the relationship among the atoms of a molecule and also handle basic information about the compounds (such as molecular weight, electrical charge, etc.). Fig. 1 shows the representation of the chemical compound TR-339 (the *2-amino-4-nitrophenol*) of the NTP data set. In this representation, there are three predicates:

- *atom(C, A, E, V)* gives information about an atom. C is the chemical compound where the atom belongs; A is the number of the atom in the chemical compound; E is the chemical element; and V is the electrical charge of the atom. For instance, atom(tr339, 1, O, -1) is the atom 1 of the compound tr339, It is an oxygen, and its charge is -1.
- *bond(C, A1, A2, B)* indicates the kind of bond between two atoms. C is the chemical compound where the bond belongs; $A1$ and $A2$ are the atoms of the compound connected by the bound; B is the kind of bond: simple, double or triple. For instance, bond(tr339, 9, 10, 1) is a simple bound of the chemical compound tr339 that connects the atoms 9 and 10.
- *atomcoord(C, A, X, Y, Z)*. It gives the spatial coordinates of the compound atoms. C is the chemical compound, A is the atom and X, Y and Z are the spatial coordinates. For instance, atomcoord(tr339, 1, 3.0918, -0.8584, 0.0066) indicates that the atom 1 of the compound tr339 has as coordinates (3.0918, -0.8584, 0.0066).

Fig. 1 represents the compound TR-339 with 17 atoms (3 oxygen, 2 nitrogen, 6 carbon and 6 hydrogen); there are double bonds between atoms 2 and 3; 5 and 6; 7 and 9; and 4 and 11 (see Fig. 1); and the rest of bonds are simple.

The representation introduced in [8] has a different approach: the compounds are organized according to their active centers (chemically identified with weak bonds). Active centers are atoms or groups of atoms responsible of the reactivity of the compound with biological receptors (for instance, toxicity). With this approach, each resulting part of the compound receives a code, therefore the chemical substances are represented as a string of codes.

The Viniti's group [8] proposed the *fragmentary code of substructure superposition* (FCSS) language, allowing the description of chemical compounds as a set of substructures containing the active centers. The elements of the FCSS language are chains of carbon pairs that begin and end with the descriptors of active centers. For instance, the chemical compound TR-339 described in FCSS is the following code: *9 6,06 0700151 0700131 0700331 1100331 0200331 0764111 0263070 0262111*.

2.1 Representation using a chemical ontology

The representation of chemical compounds we propose is the *chemical ontology* based on the terminology used by chemists, i.e the IUPAC nomenclature (`http://www.chem.qmul.ac.uk/iupac/`). Also we take into account the experience of previous research (specially the works in [17, 15, 8]) since we represent a chemical compound as a structure with substructures. Our point is that there is no need to describe in detail the properties of individual atom properties in a molecule when the domain ontology has a characterization for the type of that molecule. For instance, the *benzene* is an aromatic ring composed by six carbon atoms with some well-known properties. While using SAR models would represent a given compound as having six carbon atoms related together (forming an aromatic ring), in our approach we simply state that the compound is a benzene (abstracting away the details and properties of individual atoms).

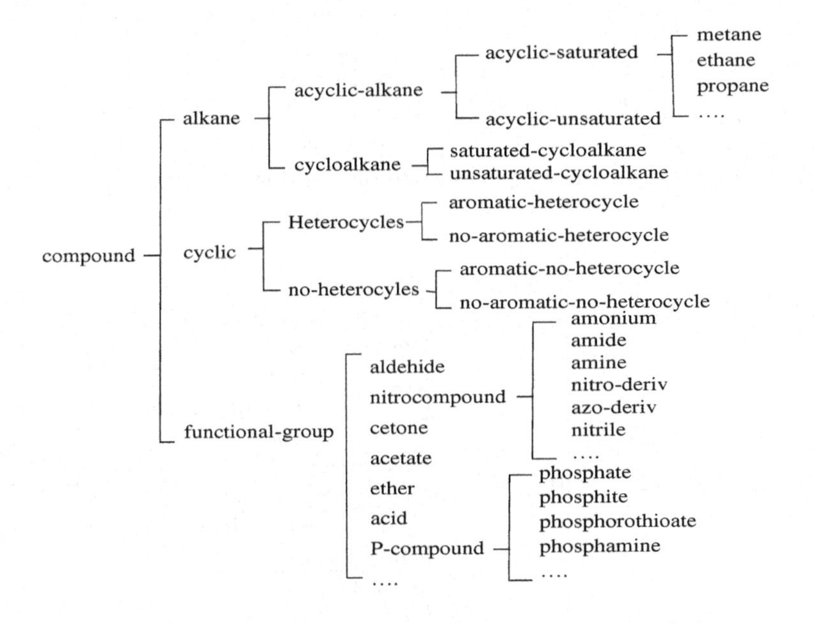

Fig. 2. Partial view of the Toxicology ontology

Fig. 2 shows a partial view of the chemical ontology we used for representing the compounds in the Toxicology data set. This ontology is based on the chemical nomenclature which, in turn, is a systematic way of describing molecules. In fact, the name of a molecule, when the standard nomenclature is used, provides to the chemist with all the information needed to graphically represent its structure. According to the chemical nomenclature rules, the name of a compound is usually formed in the following way: *radicals' names* + *main group*. Commonly, the *main group* is the part of the molecule that is either the largest or that located in a central position; however, there is no general rule to establish them. *Radicals* are groups of atoms usually smaller than the main group. A main group can have several radicals and a radical can, in turn, have a new set of radicals. Any group of atoms could be main group or radical depending on their position or relevance on the molecule, i.e., the benzene may be the main group in one compound and a radical in some other compounds.

The implementation of this representation is done using the *feature terms* formalism introduced in [1]. This formalism organizes concepts into a hierarchy of *sorts* (as that of Fig. 2), and represent descriptions and individuals as collections of features (functional relations). Sorts have an informational order relation ($\preceq$) among them, where $\psi \preceq \psi'$ means that ψ has less information than ψ' or, equivalently, that ψ is more general than ψ'. The minimal element ($\perp$) is called *any* and it represents the minimum information; when a feature value is not known it is represented as having the value *any*. All other sorts are more specific that *any*. The most general sort in Fig. 2 is *compound*. This sort has three subsorts: *alkane, cyclic* and *functional-group*,

compound	
main-group	= *compound*
p-radicals	= *position-radical*

position-radical	
position	= position
radicals	= *compound*

Fig. 3. Features corresponding to sorts *compound* and *position-radical.*

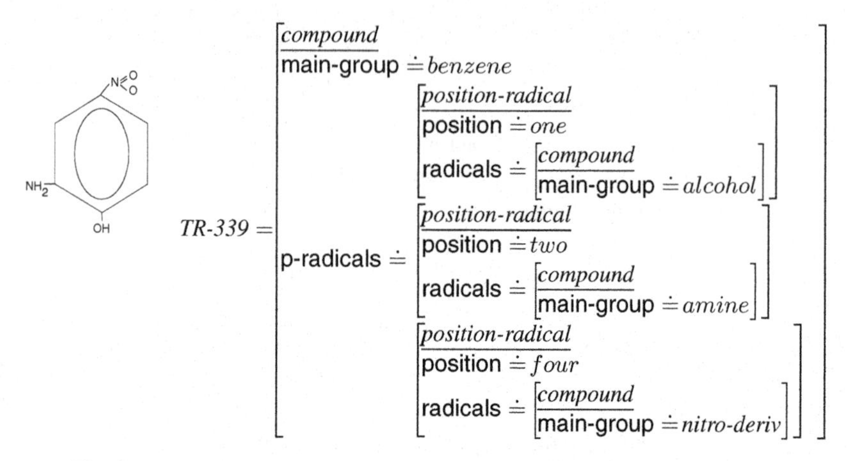

Fig. 4. Representation of TR-339, *2-amino-4-nitrophenol*, with feature terms.

which in turn, have other subsorts. The sort *methane* is more specific than the sort *acyclic-alkane*; while the sorts *methane* and *ethane* are not directly comparable.

Each sort has a collection of features characterizing the relations for this sort. For instance, Fig. 3 shows that the sort *compound* has two features: main-group and p-radicals. The values of the feature main-group have to be of the sort *compound,* while the feature p-radicals has values of sort *position-radical.* The sort *position-radical* (Fig. 3) has, in turn, two features: radicals and position. The feature radicals has values of sort *compound* (since radicals themselves are compounds). The feature position indicates where the radical(s) is bound to the main group.

Fig. 4 shows the representation of the chemical compound TR-339, *2-amino-4-nitrophenol,* using feature terms. TR-339 has a benzene as main group and a set of three radicals: an *alcohol* in position one; an *amine* in position two; and a *nitro-deriv* in position four. Notice that this information has been directly extracted from the chemical name of the compound following the nomenclature rules.

This kind of description has the advantage of being very close to the representation that an expert has of a molecule from the chemical name. We have translated, with the support of a chemist, the compounds of the NTP data set to this representation based on the chemical ontology. A shortcoming of the representation based on the chemical name of a compound is the existence of synonymous names. Currently, we have selected one of the possible names and we codified the compound with feature terms using this selected name.

3 The predictive toxicology task

The NTP data set contains reports of experiments on chemical compounds in order to establish whether they are carcinogenic. Each experiment is performed in two species: rats and mice. Moreover, because the carcinogenic activity of the compounds has proved to be different in both species and also among the sex of the same species, some computational approaches take separately the results of the experiments having, in fact, four data sets: male rats (MR), female rats (FR), male mice (MM) and female mice (FR). The chemical compounds can be classified in each data set into two solution classes: *positive* (i.e., when the compound is carcinogenic) and *negative* (i.e., when the compound is not carcinogenic).

The goal of predictive toxicology is to develop models able to predict whether a chemical compound is toxic or not. The construction of these models by computer assisted techniques takes into account the toxicity observed in some molecules to extract theories about the toxicity on families of molecules. Early systems focused on predictive toxicology were DEREK [27] and CASE [23]. DEREK is a knowledge-based system based on a set of rules describing relations between structural features and their associated toxicity. To determine the toxicity of a new compound, DEREK compares this new compound with all the compounds of the knowledge base. CASE has a base of substructures labeled as *active* or *inactive* according to their toxicity. Thus, to determine the toxicity of a new compound, CASE extracts all its possible substructures and labels each one as active or inactive using the base of substructures. Then CASE uses statistical techniques to determine the global toxicity of the new compound.

There are two families of methods currently used to solve the predictive toxicology task: statistics and ML. A widely used statistical method is regression analysis of molecular descriptors. This technique finds equations that correlate the toxicity of a compound with some physical-chemical properties [21] or with the presence of some functional groups [7]. Probabilistic reasoning such as Bayesian networks has also been widely used to build classifiers [32] or in combination with other techniques like multi-way recursive partitioning [25] and artificial neural networks [6, 5].

The focus of the second PTC [30] was to use ML to address the predictive toxicology problem. From the ML point of view, the goal of the predictive toxicology is a classification task, i.e., toxic compounds are classified as belonging to the *positive* class and non-toxic compounds are classified as belonging to the *negative* class. Moreover, the classification task has to be solved separately for each data set (MR, FR, MM and FM).

The majority of this work was concerned with using inductive techniques to construct toxicity models. Given a solution class C, a set of examples P belonging to C, and a set of examples N that do not belong to C, the goal of inductive learning techniques is to build a general description d of C such that 1) d is satisfied by all the examples in P, and 2) d is not satisfied by the examples in N.

Some inductive techniques build decision trees as predictive classifiers. The representation of the compounds is propositional (in the form of attribute value pairs) and the attributes are the values of molecular properties (molecular weight, physical-chemical properties, etc.) and results of toxicity of some other tests. The main shortcoming of decision trees is the propositional representation of the compounds due to two reasons: 1) the high number of descriptors for a compound, and 2) the fact that not all them are equally relevant in order to predict the toxicity. Most approaches use ML and statistical methods to select feature subsets.

A widely used relational learning technique is *inductive logic programming* (ILP). The main idea of ILP is to induce general descriptions explaining a set of examples represented using logical predicates. The first ILP program used to induce SAR models was PROGOL [29]; it was applied to a set of 230 aromatic and heteroaromatic nitro compounds and the resulting model was compared with models obtained by both linear regression and neural networks with backpropagation. PROGOL's results were very encouraging since the final rules were more understandable than those obtained using the other methods.

Other relational representation approaches consider a compound as a group of substructures instead of sets of atoms. These approaches consider that if a substructure has known toxic activity, then a compound having this substructure can also have toxic activity. Pfahringer and Gini proposed a more abstract representation of the chemical compounds using the concept of functional groups (similar to the chemical ontology we use, see Sect. 2.1). This abstraction improves the search process since it represents substructures rather than describing each atom and atom bonds.

Several authors [14, 17, 11] represent the compounds as labeled graphs and this allows the use of graph search algorithms for detecting frequent substructures of the molecules in the same class. Following this approach, SUBDUE [20] discovers substructures beginning with substructures matching a single vertex in the graph and extending them by selection of the best substructure in each iteration. At the end of the process, SUBDUE has a hierarchical description of the data in terms of the discovered substructures. SMILES [31], also following this approach, detects the set of molecular substructures (subgraphs) more frequently occurring in the chemical compounds.

There are also hybrid approaches, such as the one proposed by Gini et al. [16]. This approach combines the toxicity results given by a set of fragments of structures with an artificial neural network that uses descriptors of the chemical compounds. Thus, first the authors defined a set of fragments that experts recognize as structures responsible for carcinogenicity. Then they developed a module that searches in the chemical compound structure for the presence of one or more of these fragments. On the other hand, they also used an artificial neural network that assessed the carcinogenicity of a chemical compound taking into account its molecular descriptors. Finally, an ILP module is used to combine the toxicity assessment of the two modules.

A problem with inductive techniques is that the high variability of chemical compounds poses great difficulties to find general rules describing the classes appropriately. In the next section we will introduce our work on lazy learning techniques for predictive toxicology.

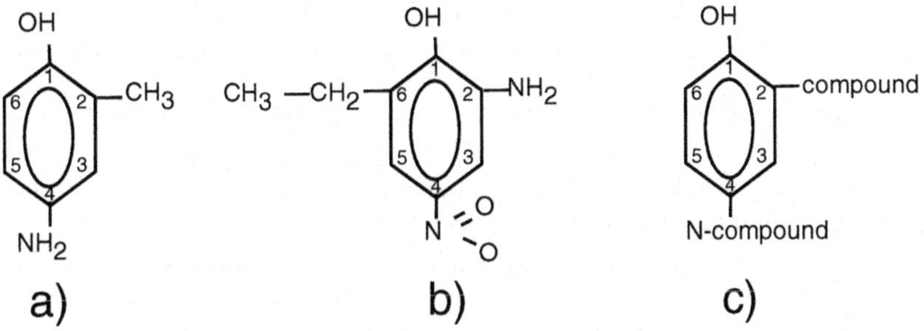

Fig. 5. a) 2-methyl-4 aminophenol. b) 2-amino-4-nitro-6-ethanophenol. c) structure shared by the chemical compounds a) and b). *compound* and *N-compound* are the most specific sort (*lub*) of the radicals in the respective positions, according to the sort/subsort hierarchy in Fig. 2

3.1 Lazy learning techniques

Inductive learning techniques try to extract general rules describing the cases in each class. This kind of techniques has some difficulties in dealing with domains, like toxicology, where entities are subject to high variability. Lazy learning techniques, on the other hand, are based on the retrieval of a set of solved problems (*cases*) similar to a specific problem. A critical issue in lazy learning is the evaluation of similarity between two cases, as this forms the basis for identifying a suitable set of cases or 'promising' candidates. Several authors use the concept of similarity between chemical compounds: HazardExpert [12] is an expert system that evaluates the similarity of two molecules based on the number of common substructures; Sello [28] also uses the concept of similarity but the representation of the compounds is based on the energy of the molecules.

Shaud

When the domain objects have a propositional representation, the similarity between two objects is assessed by computing the similarity of attributes and then aggregating their similarities to obtain a global measure of the similarity of the objects. Shaud is a similarity measure able of assessing the similarity between structured objects represented as feature terms. Given two objects Shaud distinguishes two parts in their structure: one formed by the features present in both objects, called the *shared structure*; and another formed by those features that are only present in one of the objects (but not the other) called the *unshared structure*. For instance, Fig. 5 shows that the molecules a) and b) have in common the structure c). In this example, the unshared structure is only the radical *ethane* in position six of the molecule b).

Shaud [2, 3] assesses the similarity of two feature terms by computing the similarity of the shared structure and then normalizing this value taking into account both the shared and the unshared structure. The comparison of the shared structure is performed element by element comparing the position of their sorts into the sort/subsort hierarchy in the following way:

$$S(sort(\psi^1), sort(\psi^2)) = \begin{cases} 1 & \text{if } sort(\psi^1) = sort(\psi^2) \\ 1 - \frac{1}{M} level(lub(sort(\psi^1), sort(\psi^2)) & \text{otherwise} \end{cases}$$

The idea is that the similarity between two values depends on the level of the hierarchy (see Fig. 2) where their least upper bound (*lub*) is situated in the sort hierarchy: the more general *lub*(v_1, v_2) the smaller is the similarity between v_1 and v_2. M is the maximum depth of the sort hierarchy.

For instance, in order to assess the similarity of the molecules a) and b) in Fig. 5, Shaud takes into account the structure shared by both molecules (c) and compares the elements composing that structure (Fig. 6). The similarity assessment of the shared structure is the following:

- the main group that is *benzene* in both molecules, therefore

$$S(benzene, benzene) = 1$$

- a radical in position 1 that is an alcohol in both molecules, therefore

$$S(alcohol, alcohol) = 1$$

- a radical in position 2 that is a *methane* in the molecule a) and an *amine* in the molecule b), therefore

$$S(methane, amine) = 1 - \frac{1}{M} level(lub(methane, amine))$$

and since $lub(methane, amine) = compound$, $M = 5$, and *level(compound)* = 5 (see Fig. 2) then

$$S(methane, amine) = 1 - \frac{1}{5} level(compound) = 1 - \frac{1}{5}5 = 0$$

- a radical in position 4 that is an amine in the molecule a) and a nitro-derivate (nitro-deriv) in the molecule b), therefore

$$S(amine, \text{nitro-deriv}) = 1 - \frac{1}{M} level(lub(amine, \text{nitro-deriv}))$$

and since *lub(amine, nitro-deriv)* = *N-compound*, $M = 5$ and *level(N-compound)* = 3 (see Fig. 2) then

$$S(amine, \text{nitro-deriv}) = 1 - \frac{1}{5} level(\text{N-compound}) = 1 - \frac{1}{5}3 = 0.4$$

Because these are simple molecules where the radicals themselves have no radicals, the similarity of the common part is

$$S(benzene, benzene) + S(methane, amine) + S(amine, \textit{nitro-deriv}) = 2.4$$

Then, this value is normalized by the total number of nodes (those of the shared structure plus those of the unshared structure), i.e., $S(a, b) = \frac{2.4}{5} = 0.48$.

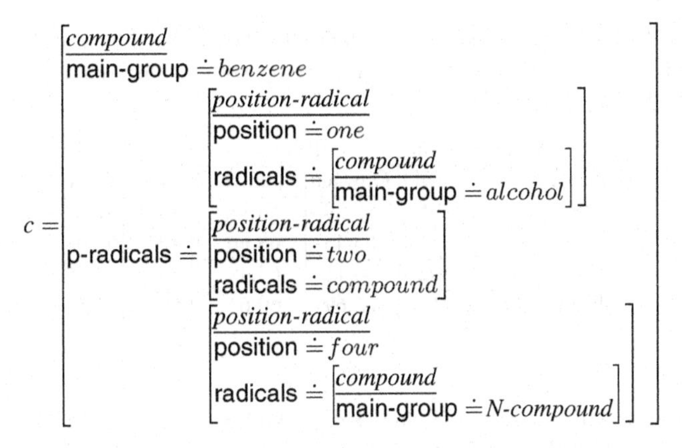

Fig. 6. Formal representation of molecule c) shown in Fig. 5

We have performed several experiments using the *k-nearest neighbor* (*k-NN*) algorithm [13]. Given a new problem p, the k-NN algorithm retrieves the k most similar cases and classifies p into the class resulting of the aggregation of the classes where the k cases belong. There are two key issues in the k-NN algorithm: the similarity measure and the aggregation. In our experiments, we took Shaud as similarity and the majority class (MC) for aggregation (i.e., the new compound is classified as belonging to the class that most of the k retrieved precedents belong to). However, our preliminary experiments using the majority criterion with different values of k did not provide a satisfactory accuracy. We proposed the *Class Similarity Average* (CSA) criterion [3], a domain-independent criterion that takes into account the similarity of the k most similar cases and also the solution class where they belong.

For each compound p to be classified, Shaud yields the similarity between p and each one of the k most similar cases. CSA will compute the average of the similarity of the cases in the same class; then the class with higher average similarity is selected as solution for p. More formally, let p be the compound to be classified and R_k the set of the k cases most similar to p according to the Shaud results. Each case $c_i \in R_k$ has the following data associated: 1) the structural similarity s_i between p and c_i, i.e., $s_i = \mathsf{Shaud}(p, c_i)$; and 2) for each data set (i.e., MR, FR, MM and FM) the compound c_i is *positive* or *negative*.

For each data set, let A^+ be the set containing cases $c_i \in R_k$ with positive activity, and A^- be the set containing cases $c_i \in R_k$ with negative activity. From the sets A^+ and A^- we define sim^+ and sim^- as the respective averages of the similarities of positive and negative cases retrieved, i.e.,

$$sim^+ = \frac{1}{|A^+|}\sum_{c_i \in A^+} s_i \text{ and } sim^- = \frac{1}{|A^-|}\sum_{c_i \in A^-} s_i$$

The carcinogenic activity of a compound c is obtained according to the following criterion (CSA): *if sim-pos < sim-neg then c has negative carcinogenic activity else c has positive carcinogenic activity.*

Table 1. Distribution of the examples in the four PTC data sets and the accuracy results obtained by two of the authors presented at the PTC compared with the accuracy of Shaud with $k = 5$ and the MC and CSA criteria.

	Composition			PTC		Acc (k = 5)	
data set	+	−	total	Ohwada	Boulicaut	MC	CSA
MR	81	125	206	55.59	55.88	48.06	**62.13**
FR	66	140	206	65.43	**68.66**	49.03	64.08
MM	63	139	202	64.11	63.49	60.40	**64.85**
FM	78	138	216	**63.69**	60.61	59.72	62.50

Table 1 shows the results of using k-NN with $k = 5$ both MC and the CSA criteria together with the accuracy of two methods presented by [26, 9] in the PTC. Notice that the accuracy using the CSA criterion is higher than using MC. Also, the accuracy taking separately positive and negative examples is more balanced using the CSA criterion. In particular, for the MR data set, the accuracies using MC are $Acc^+ = 35.80$ and $Acc^- = 56$ whereas the accuracies using CSA are $Acc^+ = 55.55$ and $Acc^- = 66.40$.

Lazy induction of descriptions

Lazy induction of descriptions (LID) is a lazy concept-learning technique for classification tasks in case-based reasoning (CBR). LID determines which are the more relevant features of a problem and searches in the case base for cases sharing these relevant features. The problem is classified when LID finds a set of relevant features shared by a subset of cases all them belonging to the same solution class C_i. Then LID classifies the problem as belonging to C_i. We call *similitude term* the structure formed by these relevant features and *discriminatory set* the set of cases satisfying the similitude term. The similitude term is a feature term composed of a set of features shared by a subset of cases belonging to the same solution class.

Given two feature terms, there are several similitude terms, LID builds the similitude term with the most relevant features. The relevance of a feature is heuristically determined using the *López de Mántaras* (LM) distance [24]. The LM distance assesses how similar two partitions are in the sense that the lesser the distance the more similar they are (see Fig. 7). Each feature f_i of an example induces a partition P_i over the case base according to the values that f_i can take in the cases. On the other hand, the LM considers the *correct partition* P_c that is the partition where all the cases contained into a partition set belong to the same solution class.

Given two partitions P_A and P_B of a set S, the distance among them is computed as follows:

$$LM(P_A, P_B) = 2 - \frac{I(P_A) + I(P_B)}{I(P_A \cap P_B)}$$

where $I(P)$ is the information of a partition P and $I(P_A \cap P_B)$ is the mutual information of two partitions.

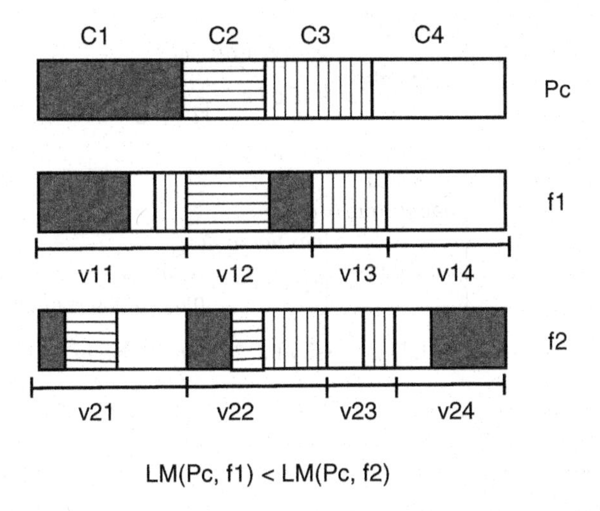

Fig. 7. Intuitive idea of the LM distance. The case base B contains precedents belonging to four solution classes. The partition induced by the feature $f1$ is more similar to the correct partition Pc than the partition induced by $f2$.

In our case, the distance measure is applied to compute the distance among a partition generated by a feature and the correct partition. The correct partition P_c has two classes, one containing the positive examples (examples in C_k) and the other containing the negative examples (those not in C_k). Thus, for each feature f_i, there is a partition P_i of the case base B according to the values of f_i. Each partition P_i is compared with the correct partition P_c using the López de Mántaras distance. The most discriminatory feature f_d is that producing a partition P_d having the minimum distance $LM(P_d, P_c)$ to the correct partition P_c.

Let P_c be the correct partition and P_i and P_j the partitions induced by features f_i and f_j respectively. We say that the feature f_i is *more discriminatory than* the feature f_j iff $LM(P_i, P_c) < LM(P_j, P_c)$. In other words, when a feature f_i is more discriminatory than another feature f_j the partition that f_i induces in B is closer to the correct partition P_c than the partition induced by f_j. Intuitively, the most discriminatory feature classifies the cases in B in a more similar way to the correct classification of cases. LID uses the *most discriminatory than* relationship to estimate the features that are most relevant for the purpose of classifying a new problem.

Now, we will illustrate the performance of LID (see algorithm in Fig. 9) to assess toxicity of the *2-chloroethanol* (the TR-275 in the PTC case base) for male rats. LID inputs are $S_D = B$ of chemical compounds, a similitude term D initialized to the most general feature term (i.e., the most general description), the description of the *2-chloroethanol*, and the set S_D (the discriminatory set associated to D) that contains

$$D^1 = \left[\begin{array}{l} \underline{\mathit{compound}} \\ \text{p-radicals} \doteq \left[\begin{array}{l} \underline{\mathit{position\text{-}radical}} \\ \text{radicals} \doteq \left[\begin{array}{l} \underline{\mathit{compound}} \\ \text{main-group} \doteq \mathit{alcohol} \end{array}\right] \end{array}\right] \end{array}\right]$$

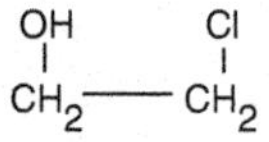

OH Cl
CH_2 — CH_2
2-chloroethanol

$$D^2 = \left[\begin{array}{l} \underline{\mathit{compound}} \\ \text{main-group} \doteq \mathit{ethane} \\ \text{p-radicals} \doteq \left[\begin{array}{l} \underline{\mathit{position\text{-}radical}} \\ \text{radicals} \doteq \left[\begin{array}{l} \underline{\mathit{compound}} \\ \text{main-group} \doteq \mathit{alcohol} \end{array}\right] \end{array}\right] \end{array}\right]$$

Fig. 8. Similitude terms build by LID to classify the *2-chloroethanol* in the negative class for male rats.

```
Function LID (Δ_D, p, D, C)
    if stopping-condition(Δ_D)
       then return class(Δ_D)
       else f_d := Select-feature (p, Δ_D, C)
            D' := Add-feature(f_d, D)
            Δ_D' := Discriminatory-set (D', Δ_D)
            LID (Δ_D', p, D', C)
    end-if
end-function
```

Fig. 9. The LID algorithm. D is the similitude term, Δ_D is the discriminatory set of D, C is the set of solution classes, $class(\Delta_D)$ is the class $C_i \in C$ to which all elements in Δ_D belong.

all the cases that satisfy the structure described by D. Initially $S_D = B$ since D is satisfied by all the cases in B.

The first step of LID is to check whether all the cases in Δ_D belong to the same solution class. Since this stopping condition is not satisfied at the beginning, the second step is to specialize D. The specialization D^1 of D is built by adding to D the path p-radicals.radicals.main-group with main-group taking value *alcohol*, as in the *2-chloroethanol* (see Fig. 8). The discriminatory set Δ_{D_1} contains now 42 cases subsumed by D^1, i.e., those compounds in Δ_D having a radical alcohol. Next, LID is recursively called with D^1 and Δ_{D_1}.

The cases in the discriminatory set Δ_{D_1} do not satisfy the stopping condition, i.e., some of them belong the positive class and some others belong to the *negative* class, therefore D^1 has to be specialized by adding a new discriminatory feature. Now most discriminatory feature is main-group. The specialization D^2 is built by adding main-group to D^1 with value *ethane* (see Fig. 8). LID is recursively called with the set Δ_{D_2} and the similitude term D^2.

The set Δ_{D_2} contains 6 cases all of them belonging to the negative class. Therefore LID terminates classifying the *chloroethanol* as belonging to the *negative* class and explaining it with the similitude term D^2 (shown in Fig. 8), i.e., because the compound is an ethane with a radical alcohol. This classification is supported by the 6 cases in Δ_{D_2}. The result of LID is the solution class C_i and a similitude term D^n. The similitude term D^n can be seen as an explanation of why the current problem p is in the solution class C_i. D^n is a *partial* description of C_i because, in general, not all cases in C_i satisfy D^n.

We conducted a series of experiments with the similitude terms to discover patterns in the Toxicology data set. These experiment had two steps: 1) use LID with the leave-one-out method in order to generate similitude terms for classifying the cases; and 2) select a subset of these similitude terms. The first step yields a set of similitude terms that have been used for classifying some cases. The second step selects only those similitude terms that are totally discriminatory (we call them *patterns*). Some of the patterns detecting positive toxicity are also reported in the literature. For instance, LID founds that compounds with a radical chlorine are carcinogenic and Brautbar describes some experiments confirming the toxicity of chlorinated hydrocarbons.

As a second experiment, we defined Caching LID (C-LID), a lazy learning approach that reuses the patterns used for solving past problems in order to improve the classification of new problems in case-based reasoning (CBR). C-LID is implemented on top of LID by defining two policies: the caching policy and the reuse policy. The *caching policy* determines which similitude terms (patterns) are to be retained. The *reuse policy* determines when and how the cached patterns are used to solve new problems. In our experiments, the caching policy of C-LID states that a similitude term D will be cached if it is *univocal*, i.e., when all cases covered by a pattern belong to one class only. The *reuse policy* of C-LID states that patterns will be used for solving a problem p only when LID is unable to univocally classify p.

Thus, the experiment with C-LID has two phases: 1) a preprocessing of the case base in order to obtain some patterns to be cached; and 2) the problem solving phase that uses LID together with the cached patterns for classifying new problems. The preprocessing phase is done using the leave-one-out technique using the cases in the case base B. For each case $c \in B$, C-LID uses LID to classify c and generates a similitude term D_c. When D_c is univocal C-LID caches it. Thus, at the end of the preprocessing phase C-LID has obtained a set $M = \{D_1 \ldots D_n\}$ of patterns. The reuse policy decides when to use these patterns during the problem solving phase.

The evaluation of the predictive accuracy of the methods has been made using 10-fold cross-validation. Table 2 shows the accuracy of LID and C-LID for each one of the data sets. Notice that C-LID improves the accuracy of LID in all the data sets showing that the caching policy is adequate. Notice that the caching policy stores only the similitude terms that are univocal, i.e., those subsuming cases belonging to only one solution class. With this policy C-LID takes into account only those patterns with clear evidence of a good discrimination among classes.

Table 2. Accuracy of LID and C-LID on the PTC data set.

data set	# cases	LID	C-LID
MR	297	58.27	60.54
FR	296	63.09	66.97
MM	296	52.39	53.95
FM	319	52.36	56.60

4 Conclusions

We have seen that the task of predicting the possible activity of molecules is a challenging one, from the chemist viewpoint and also the field of ML. From the chemist viewpoint it is interesting that automated techniques may be capable of predicting with some degree of accuracy the toxicity of chemical compounds that have not been synthesized. Predicting toxicity is a complex task for ML that requires thoughtful analysis of all dimensions involved.

We have summarily described several ML approaches to toxicity prediction, and we have highlighted the dimension of example representation. ML approaches that use a propositional representation (i.e., an example is represented by a vector of attribute value pairs) have problems for mapping the chemical model of chemical compounds based on SAR into vectors of attribute value pairs. Since this mapping ignores the structure itself, other ML approaches use relational learning techniques; specifically ILP maps the SAR models into a logic representation of examples and background knowledge. Our approach proposes a new kind of relational representation based on the chemical ontology that describes the compounds' structure in a more abstract way. The experiments have shown that the predictive performance of our methods (SHAUD and C-LID using the chemical ontology based representation) have comparable results to that of methods that use SAR models.

ML techniques are very dependent on the way examples are represented. The fact that ML techniques—using propositional SAR, relational SAR, and chemical ontology—achieve a similar performance in predicting toxicity implies that they possess a comparable information content in terms of the studied molecules. Nonetheless, toxicity prediction is a complex task for ML techniques, since their performance is just relatively good [19] while they can be very good for other tasks. Because there is no ML technique providing excellent results, a likely explanation is that the current representation of chemical compounds is not adequate. Notice that a compound can be toxic in a data set (say male rats) and not in another (say female mouse): since the representation of the examples is the same, and yet they have different solutions, this seems to indicate that there are external factors involved that are not represented in the examples themselves. An enriched characterization of the compounds would very likely improve the predictive accuracy of the ML techniques we have discussed here.

Acknowledgments

This work has been supported by the SAMAP project (TIC2002-04146-C05-01). The authors thank Josep Lluís Arcos and Lluís Bonamusa for their support in the elaboration of this paper.

References

1. E. Armengol and E. Plaza. Lazy induction of descriptions for relational case-based learning. In L. De Reaedt and P. Flach, editors, *ECML-2001. Freiburg. Germany.*, number 2167 in Lecture Notes in Artificial Intelligence, pages 13–24. Springer, 2001.
2. E. Armengol and E. Plaza. Similarity assessment for relational cbr. In David W. Aha and Ian Watson, editors, *CBR Research and Development. Proceedings of the ICCBR 2001. Vancouver, BC, Canada.*, number 2080 in Lecture Notes in Artificial Intelligence, pages 44–58. Springer-Verlag, 2001.
3. E. Armengol and E. Plaza. Relational case-based reasoning for carcinogenic activity prediction. *Artificial Intelligence Review*, 20(1–2):121–141, 2003.
4. J. Ashby and R.W. Tennant. Prediction of rodent carcinogenicity for 44 chemicals: results. *Mutagenesis*, 9:7–15, 1994.
5. D. Bahler, B. Stone, C. Wellington, and D.W. Bristol. Symbolic, neural, and bayesian machine learning models for predicting carcinogenicity of chemical compounds. *J. of Chemical Information and Computer Sciences*, 8:906–914, 2000.
6. S.C. Basak, B.D. Gute, G.D. Grunwald, D.W. Opitz, and K. Balasubramanian. Use of statistical and neural net methods in predicting toxicity of chemicals: a hierarchical qsar approach. In G.C. Gini and A.R. Katrizky, editors, *Predictive Toxicology of Chemicals: Experiences and Impacts of AI Tools*, pages 108–111. AAAI Press, 1999.
7. E. Benfenati, S. Pelagatti, P. Grasso, and G. Gini. Comet: the approach of a project in evaluating toxicity. In G.C. Gini and A.R. Katrizky, editors, *Predictive Toxicology of Chemicals: Experiences and Impacts of AI Tools*, pages 40–43. AAAI Press, 1999.
8. V. Blinova, D.A. Bobryinin, V.K. Finn, S.O. Kuznetsov, and E.S. Pankratova. Toxicology analysis by means of simple jsm method. *Bioinformatics*, 19(10):1201–1207, 2003.
9. J.F. Boulicaut and B. Cremilleux. δ-strong classification rules for characterizing chemical carcinogens. In *Proceedings of the Predictive Toxicology Challenge Workshop, Freiburg, Germany, 2001.*, 2001.
10. D.W. Bristol, J.T. Wachsman, and A. Greenwell. The niehs predictive toxicology evaluation project. *Environmental Health perspectives*, 104:1001–1010, 1996.
11. R. Chittimoori, L. Holder, and D. Cook. Applying the subdue substructure discovery system to the chemical toxicity domain. In *Proceedings of the Twelfth International Florida AI Research Society Conference, 1999*, pages 90–94, 1999.
12. F. Darvas, A. Papp, A. Allerdyce, E. Benfenati, and G. Gini et al. Overview of different ai approaches combined with a deductive logic-based expert system for predicting chemical toxicity. In G.C. Gini and A.R. Katrizky, editors, *Predictive Toxicology of Chemicals: Experiences and Impacts of AI Tools*, pages 94–99. AAAI Press, 1999.
13. B.V. Dasarathy. *Nearest neighbor (NN) norms: NN pattern classification techniques*. Washington; Brussels; Tokyo; IEEE computer Society Press, 1990.
14. L. Dehaspe, H. Toivonen, and R.D. King. Finding frequent substructures in chemical compounds. In R. Agrawal, P. Stolorz, and G. Piatetsky-Shapiro, editors, *4th Int. Conf. on Knowledge Discovery and Data Mining*, pages 30–36. AAAI Press., 1998.

15. M. Deshpande and G. Karypis. Automated approaches for classifying structures. In *Proc. of the 2nd Workshop on Data Mining in Bioinformatics.*, 2002.
16. G. Gini, M. Lorenzini, E. Benfenati, R. Brambilla, and L. Malvé. Mixing a symbolic and subsymbolic expert to improve carcinogenicity prediction of aromatic compounds. In *Multiple classifier systems. 2th Intern. Workshop*, pages 126–135, 2001.
17. J. Gonzalez, L. Holder, and D. Cook. Graph based concept learning. In *AAAI*, page 1072, 2000.
18. C. Helma, E. Gottmann, and S. Kramer. Knowledge discovery and data mining in toxicology. *Statistical Methods in Medical Research*, 9:329–358, 2000.
19. C. Helma and S. Kramer. A survey of the predictive toxicology challenge 2000-2001. *Bioinformatics*, pages 1179–1200, 2003.
20. L.B. Holder, D.J. Cook, and S. Djoko. Substructure discovery in the subdue system. In *Proceedings of the AAAI Workshop on Knowledge Discovery in Databases*, pages 169–180, 1994.
21. M. Karelson and U. Maran. Qspr and qsar models derived with codessa multipurpose statistical analysis software. In G.C. Gini and A.R. Katrizky, editors, *Predictive Toxicology of Chemicals: Experiences and Impacts of AI Tools*, pages 12–23. AAAI Press, 1999.
22. A.R Katritzky, R. Petrukhin, H. Yang, and M. Karelson. *CODESSA PRO. User's manual.* University of Florida, 2002.
23. G. Klopman. Artificial intelligence approach to structure-activity studies: Computer automated structure evaluation of biological activity of organic molecules. *Journal of the America Chemical society*, 106:7315–7321, 1984.
24. R. López de Mántaras. A distance-based attribute selection measure for decision tree induction. *Machine Learning*, 6:81–92, 1991.
25. K.L. Mello and S.D. Brown. Combining recursive partitioning and uncertain reasoning for data exploration and characteristic prediction. In G.C. Gini and A.R. Katrizky, editors, *Predictive Toxicology of Chemicals: Experiences and Impacts of AI Tools*, pages 119–122. AAAI Press, 1999.
26. H. Ohwada, M. Koyama, and Y. Hoken. Ilp-based rule induction for predicting carcinogenicity. In *Proceedings of the Predictive Toxicology Challenge Workshop, Freiburg, Germany, 2001.*, 2001.
27. D.M. Sanderson and C.G. Earnshaw. Computer prediction of possible toxic action from chemical structure: the derek system. *Human and Experimental Toxicology*, 10:261–273, 1991.
28. G. Sello. Similarity, diversity and the comparison of molecular structures. In G.C. Gini and A.R. Katrizky, editors, *Predictive Toxicology of Chemicals: Experiences and Impacts of AI Tools*, pages 36–39. AAAI Press, 1999.
29. A. Srinivasan, S. Muggleton, R.D. King, and M.J. Sternberg. Mutagenesis: Ilp experiments in a non-determinate biological domain. In *Proceedings of the Fourth Inductive Logic Programming Workshop*, 1994.
30. S. Srinivasan, S.H. Muggleton, R.D. King, and M.J. Stenberg. The predictive toxicology evaluation challenge. In *IJCAI, Nagoya, Japan*, pages 4–9. Morgan Kaufman, 1997.
31. D.J. Weininger. Smiles a chemical language and information system. *J. Chem. Inf. Comput. Sci.*, 28(1):31–36, 1988.
32. C.A. Wellington and D.R. Bahler. Predicting rodent carcinogenicity by learning bayesian classifiers. In G.C. Gini and A.R. Katrizky, editors, *Predictive Toxicology of Chemicals: Experiences and Impacts of AI Tools*, pages 131–134. AAAI Press, 1999.

QSAR Modeling of Mutagenicity on Non-Congeneric Sets of Organic Compounds

Uko Maran and Sulev Sild

Department of Chemistry, University of Tartu, Tartu 51014, Estonia.
E-mail: uko.maran@ut.ee or sulev.sild@ut.ee

Summary. This chapter describes the interdisciplinary research where complex biochemical interactions of chemicals with the DNA is modeled with the aid of methods from artificial intelligence, quantum mechanics, statistical methods by analyzing relationships between the mutagenic activity of compounds and their structure. The overview is given on the use of artificial intelligence methods for the estimation of mutagenicity. The focus is on *quantitative structure-activity relationships*, the selection of molecular descriptors for the relationships and the efforts of modeling described in the literature.

1 Introduction

The reliable assessment of hazards posed by chemicals has a major significance when examining the potential genotoxicity of newly designed chemicals and materials. Various experimental methods have been used to measure the genotoxicity of compounds. However, most of them are expensive in terms of time and the experimental resources required [33]. Therefore, various cost- and time-efficient theoretical methods [7, 10, 23] have been developed that enable the prediction of genotoxicity at the level of reasonable experimental quality. These methods originate from Hansch's work in 1960s [22], where he linked the biochemical activity of chemicals and the experimentally derived parameters (descriptors) capturing hydrophobic, electronic and steric interactions influencing the activity of those chemicals into one mathematical equation. By definition, each Hansch type (quantitative) structure-activity relationship is valid within a congeneric set of compounds. Congeneric compounds belong to a well-defined family of molecular structures characterized by certain (one or several) functionalities that are responsible for the observed activity. Thus, each compound from a congeneric set is assumed to produce certain biological action through the same mechanism. Different substituents attached to a common molecular skeleton simply modulate the quantitative level of activity. Consequently, the *quantitative structure-activity relationship* (QSAR) analysis is frequently aimed at developing a mathematical model that relates the biological activity to small variations of the chemical structure parameterized by empirical physico-chemical descriptors or theoretical molecular descriptors. The congeneric set of compounds is often collected

W. Dubitzky and F. Azuaje (eds.), Artificial Intelligence Methods and Tools for Systems Biology, 19–35.

using only the similarity of chemical structures, and QSAR models are developed by assuming that they will also have the same mechanism of toxic action. Therefore, it is critical how congeneric series are tailored, and this strongly determines the success of the analysis. Much work has been done in the field of QSAR analysis of congeneric sets of genotoxic compounds and several reviews [10, 45, 8, 9] describe the achievements in this field.

In risk assessment, however, one has often to deal with a large number of chemicals that are structurally diverse i.e., non-congeneric. The analysis of diverse chemicals is extremely difficult because they may follow different mechanisms of toxic action. This makes it very hard to summarize all the possible mechanisms into one model or to find some underlying factor(s) that is common to all of them. Despite of such difficulties, attempts have been made to establish predictive models for non-congeneric series of genotoxic chemicals [45].

2 Quantitative structure-property/activity relationship

A quantitative structure-property relationship (OSPR) or a quantitative structure-activity relationship describes a mathematical relationship where the property or activity (P) is a function (see equation 1) of the molecular structure that is described through the descriptors (d_i).

$$P = f(d_1, d_2, \ldots, d_n) \tag{1}$$

The dependent variable in this relationship is a physico-chemical property or a bio-chemical activity. The property can be any measured value that describes physico-chemical interactions, like boiling point, melting point, vapor pressure, partition coefficients, etc. The activity can be any value describing bio-chemical interactions, for instance, inhibitor constant, receptor binding energy, bio-concentration, bio-degradation, rate of chemical reaction, acute toxicity, mutagenicity, carcinogencity, etc.

In the QSPR/QSAR approach, molecules can be represented by a wide variety of (theoretical) molecular descriptors [28, 55] which are used as independent variables in the model. Thousands of different descriptors are currently available and they are traditionally divided into several subclasses according to the information they capture (Fig. 1):

- *Constitutional descriptors* capture information about the chemical composition of compounds. The examples are counts of atoms, bonds, functional groups, etc. Constitutional descriptors characterize the one-dimensional (1D) properties of molecules, where the chemical formula is sufficient to calculate the descriptors.
- *Topological descriptors* are numerical descriptors that are derived from the two-dimensional (2D) structure of molecules. These descriptors require information about the connectivity in molecules that is usually expressed in the form of mathematical objects—*graphs*. A topological descriptor reduces the molecular graph into a number that characterizes the structure and the branching pattern

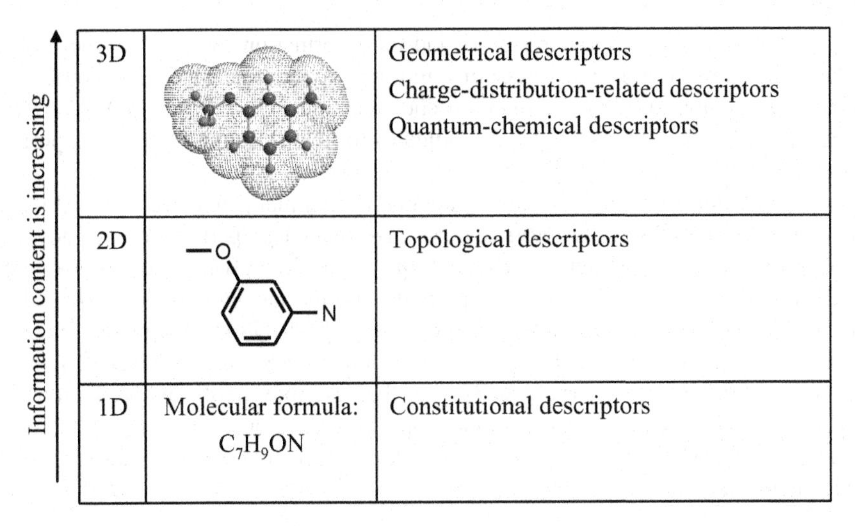

Fig. 1. Classification of descriptors according to the structural information captured

of the molecule. The common representatives of topological descriptors are various connectivity indexes. They can also include information about the nature of atoms, bond multiplicity, stereo-chemical features, and electronic parameters associated with various atoms.

- *Geometrical descriptors* are derived from the information about the orientation of atoms in space. They are calculated from the three-dimensional (3D) coordinates of atomic nuclei in space, atomic masses and/or atomic radii. The typical examples are molecular surface areas, solvent accessible molecular surface areas, moments of inertia of a molecule, etc.
- *Charge-distribution-related descriptors* combine the 3D coordinates and the information about the electronic structure of molecules. The electronic structure is particularly important because the electron densities and charges in molecules determine the physico-chemical properties (polar interactions) and reactivity of chemicals (covalent interactions). The charge distribution in the molecules can be calculated at various levels of theory starting with empirical methods and ending with quantum chemical methods. Examples of the charge-distribution-related descriptors are atomic partial charges of the atoms, charged partial surface areas, etc.
- *Quantum-chemical descriptors* are based on the molecular quantum mechanical calculations that solve the time-independent Schrdinger equation for the stationary states of molecules [36]. Quantum mechanical calculations can range from various semi-empirical approximations to a wide selection of ab initio methods. The real application of quantum mechanical calculations depends on available computer resources and the number of chemicals in the training set. From the vast amount of quantum chemical descriptors the best examples are the energies of the highest occupied molecular orbital and the lowest unoccupied molecular orbital.

The complexity of the descriptors and the information captured by the molecular descriptors rises in the rows constitutional → topological → geometrical → charge-distribution-related → quantum chemical and 1D → 2D → 3D. Therefore more complex descriptors are good candidates for the modeling of complex properties or activities.

Various statistical methods can be applied to find mathematical relationships or models between these descriptors and the investigated properties or activities (see Fig. 2). For the development of a model, the molecule must have a corresponding experimentally measured activity or property and the molecular descriptors calculated, as briefly overviewed above. The following QSPR/QSAR treatment will apply *multi-linear regression* (MLR), *partial least squares* (PLS), *artificial neural network* (ANN) or other methods for the analysis of the descriptor pool. The prediction of a property or activity involves the application of the developed QSPR/QSAR model. The QSPR/QSAR predictions have an important role in drug design, material design, and chemical engineering.

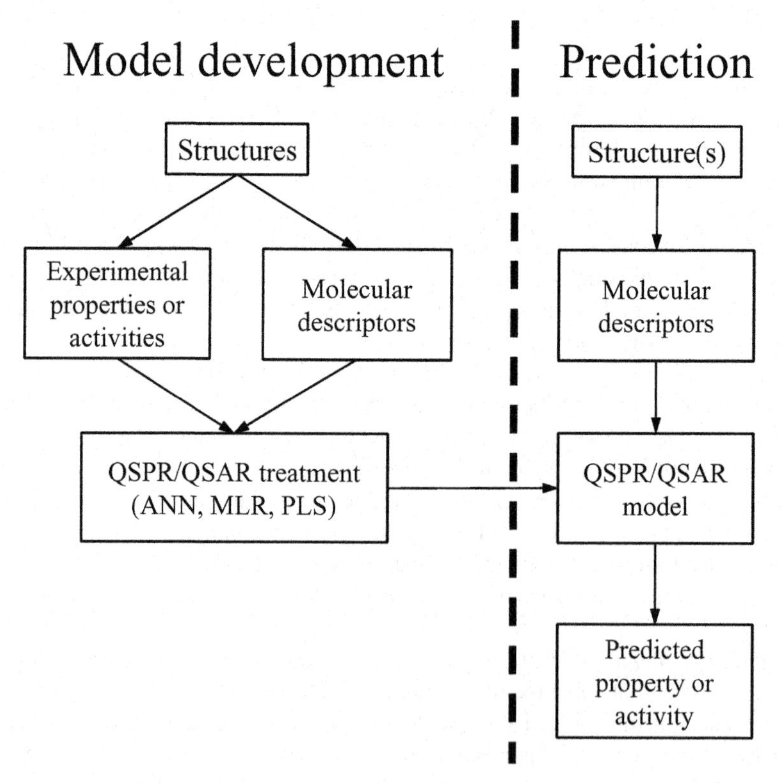

Fig. 2. Overview of quantitative structure-property or structure-activity relationship (QSPR/QSAR) approach

The developed QSPR/QSAR models (see equation 1) can be either linear or non-linear in their representation. In practice, linear models are more common, as they are much easier do develop and interpret. Non-linear models are often developed with the help of artificial intelligence (AI) methods that are getting increasingly more popular.

The quality of QSPR/QSAR models is measured by several criteria characterizing the regression models, like the *correlation coefficient* of the regression (R), *squared correlation coefficient* (R^2—the coefficient of determination), the *standard error* of the multiple linear regression (s), the *normalized standard error* of the regression (s_0), the *Fisher criterion* (F), the *Student's t-test*, the *root-mean-square* (RMS) error, etc. [28] The best QSPR/QSAR model is usually selected by analyzing a complex of different criteria.

3 Variable selection

In the case of large numbers of (experimental or theoretical) molecular descriptors the correct selection of relevant descriptors (variables) into the QSPR/QSAR models becomes important. The variable selection is a particularly important and challenging problem in the development of artificial neural network (ANN) models.

Over the years, a vast number of methods for the dimensionality reduction of the descriptor space have been suggested for the QSPR/QSAR modeling. Variable reduction is needed to remove redundancies from a descriptor pool and to eliminate molecular descriptors that contain irrelevant information about the modeled property or activity. This effectively makes QSPR/QSAR models much easier to interpret and reduces the risk for chance correlations. Traditional variable selection methods include algorithms for *forward selection* and *backward elimination* of descriptor scales, *principal component analysis* (PCA), and PLS. Also, various stochastic methods exploit evolutionary algorithms, such as *genetic algorithms* (GA) and *simulated annealing* (SA). There is no guarantee that these methods will find the most optimal subset of descriptors, although in practice, these methods frequently provide reliable and predictive correlation models. In principle, a complete analysis of all possible descriptor subsets could be performed, but this becomes impractical very quickly as the number of the available descriptor scales and the size of the subset increases.

All these methods have been successfully applied for the development of multi-linear (ML) QSPR/QSAR models for various physico-chemical properties and biological activities, as summarized in the reviews [28, 32, 29, 31]. On the other hand, the range of available descriptor selection methods for the development of ANN models has been more limited in the past due to the lack of available computer resources. Fortunately, the rapid increase of hardware performance in recent years has alleviated this situation to some extent.

The PCA method is often used in ANN modeling for dimensionality reduction. The principal component scores are extracted from the input data matrix and they are used as input variables for the ANN models. Examples include carcinogenicity [18], calcium channel antagonists [56], and the activity of bradykinin potentiating pentapeptides [57]. The risk associated with the PCA is the potential elimination of input variables that have a non-linear relationship with the dependent variables. Another problem of the principal component scores is that they are much harder to interpret, since they do not have a clear physical meaning.

The forward selection of descriptor scales is very popular for ML QSPR/ QSAR analysis, since the resulting models have a very simple mathematical formulation and are very easy to interpret. Several variations of forward selection algorithms exist [14, 51]. These methods start usually with the elimination of insignificant descriptor scales according to various criteria (e.g., too little variation in descriptor values, too many missing descriptor values, high inter-correlation with some other descriptor, etc.). Next, statistical parameters of all one- or two-parameter correlation models are evaluated, the best models are selected, and new descriptors are incrementally added. This process proceeds until QSPR/QSAR models with desired quality of prediction are obtained.

The forward selection algorithm still involves a significant number of model evaluations with different descriptor subsets, and thus, this approach is rarely used for the development ANN models, because the training process can be quite time-consuming. Since the model evaluation for ML models is much cheaper, the descriptor subsets selected with forward selection ML regression analysis have been used previously as inputs for ANNs [11]. Of course, this approach is limited to data sets where non-linear relationships with the investigated property or activity is not very significant. However, descriptor subsets from ML regression models can be a starting point for the selection of feasible ANN architectures, which follows with interactive improvement of selected subsets by the removal or addition of descriptors [30], or further optimization by genetic algorithms [42]. The forward selection with ANN modeling has been used for the development of predictive QSPR models for dielectric constants [49] and mutagenicity [38] that will be discussed in Section 5.

The backward elimination is another approach that is used in combination with ANNs to reduce the dimensionality of data. These methods are usually based on the pruning of input unit weights. The sensitivity of input units is analyzed and the least significant units are eliminated. Tetko et al. [53, 35] published several examples of different pruning algorithms for the back propagation and cascade correlation learning algorithms. Pruning methods were also used to develop the ANN models for octanol water partition coefficients [26] and drug transfer into human breast milk [2].

The GA method [19, 46] uses the principle of natural evolution for solving various optimization problems and is commonly used in connection with ANN modeling. The GA algorithm starts the descriptor selection with a population of randomly constructed QSPR/QSAR models. These models are ranked using some fitness function that evaluates the predictive power of the model. Then a set of genetic operations (crossover, mutation) is applied to the better-ranked models in the population to produce new models that will replace the worst ranked models. Thus, in analogy with the evolution theory, only the fittest models will survive. Karplus and coworkers [50] have tested different methods (forward selection with ML regression, genetic function approximation, GA-ANN, SA-ANN) to build QSAR models on progesterone receptor binding steroids. They concluded that non-linear models outperformed linear models, while the best results were obtained with the GA-ANN method. Jurs et al. have used GAs and SAs to build ANN models for auto ignition temperatures [42], boiling points [20], and the inhibition concentration of acyl-CoA:cholesterol O-acyltransferase inhibitors [43]. Other examples include human intestinal absorption [1], and the treatment of estrogen receptor ligands, carbonic anhydrase II inhibitors, and monoamine oxidase inhibitors [16].

4 Ames test of mutagenicity

The genotoxicity is a specific adverse effect on the genome of living cells that, upon the duplication of the affected cell, can be expressed as a mutagenic or a carcinogenic event because of specific alterations of the molecular structure of the genome through the translation of proto-oncogenes [25]. It results from a reaction with the DNA that can be measured either bio-chemically or, in short-term tests that reflect the DNA damage.

The genotoxicity is usually detected by the ability of a chemical to produce tumors on laboratory animals. This kind of experiment may last for two or three years and require the use of significant resources and expertise. This has led to the development of alternative, short-term, and relatively inexpensive assay screening tests for the toxic effects on the genome without the use of live animals. The most commonly used tests are for the detection of mutagenicity [27]. The best known of them is the Ames test that is based on the *Salmonella typhimurium* bacterial strain [40, 3] .

During this test, the colonies of bacteria having a reversible mutant gene are exposed to a chemical under testing [48]. This mutant gene makes the bacterial strain unable to synthesize one particular amino acid (histidine) that is essential for the bacteria to live. The testing environment contains only a limited amount of histidine which is insufficient for the normal growth. If the chemical has the mutagenic activity then it will cause mutations that will reverse the effect of the original mutation on the bacteria. Because the mutant bacteria reverse back to their original character, they are called *revertants*. After the reverse mutation the bacteria no longer require histidine from the environment as they can produce histamine again by themselves and grow normally. Thus this test measures the net revertant mutant colonies obtained at several doses of test chemical and is expressed by the logarithm of the number of revertants per nanomole. It can be measured for different bacterial strains (for instance TA98 and TA100) of *Salmonella typhimurium*, with or without microsomial preparation (+S9). Many chemicals are non-mutagenic, but are converted into mutagens as they are metabolized. The microsomial preparation adds metabolizing enzymes that are extracted from the liver cells of mammals (normally rats). The microsomial preparation is designed to simulate mammalian liver enzyme systems and is used to detect chemicals, which undergo the metabolic activation from non-mutagenic forms to mutagenic forms.

Although the Ames test allows a relatively quick and inexpensive way for detecting chemicals that are potential mutagens, the test is still far too expensive and time-consuming if a large number of chemicals have to be tested. Therefore theoretical models are useful to make the computer simulation of Ames tests possible.

5 QSAR models for mutagenicity

TOPKAT and CASE-MULTICASE [10] are programs originating from the Hansch approach [22] that have been used to predict toxicity (including genotoxicity) in non-congeneric databases. The TOPKAT (TOxicity Prediction by Komputer Assisted Technology) system is based on the quantitative structure toxicity relationships extending the classical Hansch approach to modeling of non-con-generic data by relying on the use of 2D topological and substructure-specific parameters [15]. The CASE/MULTICASE system performs a statistical analysis to identify those molecular fragments that are relevant to the observed activity [34]. The structure-activity relationship approaches (including TOPKAT) start with the analysis of the training set of chemicals, searching for the predefined set of substructural fragments, find their contribution to the modeled activity and then use the fragments with the largest contribution in the formation of the model. The TOPKAT and CASE-MULTICASE applications on non-congeneric databases have been well documented in several reviews [45, 8].

The mutagenicity has been modeled for non-congeneric sets of compounds using various variable selection algorithms combined with ML and ANN methods for developing QSAR models (Table 1).

Based on a set of 43 aminoazobenzene derivates, Garg et al. [17] have derived a five-parameter ML QSAR model (the squared correlation coefficient, $R^2 = 0.85$) and a three-parameter ANN model with the squared correlation coefficient $R^2 = 0.88$ for mutagenicity (Table 1: row #1 & row #2). Recently this work has been extended to the mutagenicity of a set of 74 aminoazo dyes [52]. The derived 8 parameter ML QSAR model was characterized with the squared correlation coefficient $R^2 = 0.73$, while the respective ANN QSAR model with 8 parameters produced a remarkably improved squared correlation coefficient $R^2 = 0.95$ (Table 1: #3 & #4).

Gramatica et al. [21] have used genetic algorithms for the variable selection from a large pool of molecular descriptors for the modeling of aromatic and heteroaromatic amines for two bacterial strains TA98 and TA100. At first they used distance-based experimental design [54] for the selection of training and validation sets. The models (Table 1: #5 & #6) were developed for both TA98 and TA100 strains, incorporating four and three descriptors, respectively. The descriptors in the models were analyzed to understand the mechanism of mutagenicity. They found that steric factors appear more important in TA98 models, while polarizability, electronic, and hydrogen bonding related factors were more prominent in TA100 models. The prediction quality of both models was assessed with various validation techniques.

Cash used electro-topological state indexes for modeling the mutagenicity of aromatic and heteroaromatic amines [12]. Stepwise forward selection was used to construct QSAR models. The best model for a data set of 95 structures had the squared

Table 1. Characteristics and Methods for QSAR models in literature: number of compounds in the data set (N), number of descriptors in model (N_D), squared correlation coefficient (R^2), squared standard error (s^2).

#	Dataset	N	Method[a]	N_D	R^2	s^2	Reference
1	Aminoazo dyes (TA98 +S9)	43	FS-ML	5	0.85	0.13	[17]
2	Aminoazo dyes (TA98 +S9)	43	n/a-ANN	3	0.88	–	[17]
3	Aminoazo dyes (TA98 +S9)	74	FS-ML	8	0.73	0.23	[52]
4	Aminoazo dyes (TA98 +S9)	74	n/a-ANN	8	0.95	–	[52]
5	Aromatic amines (TA98 +S9)	60	GA-ML	4	0.80	0.68	[21]
6	Aromatic amines (TA100 +S9)	46	GA-ML	3	0.81	0.34	[21]
7	Aromatic amines (TA98 +S9)	95	FS-ML	9	0.77	0.96	[12]
8	Aromatic amines (TA98 +S9)	95	C-ML	8	0.79	0.83	[6]
9	Aromatic amines (TA98 +S9)	95	C-ML	9	0.82	0.71	[4]
10	Aromatic amines (TA98 +S9)	90	GA-ANN	26	0.64	–	[39]
11	Aromatic amines (TA98 +S9)	95	FS-ML	6	0.83	0.66	[37]
12	Aromatic amines (TA98 +S9)	95	FS-ANN	6	0.90	–	[30]
13	Diverse set (TA100)	117	FS-ML	12	0.83	0.74	[38]
14	Diverse set (TA100)	117	FS-ANN	7	0.84	–	[38]
15	Diverse set (TA98)	211	FS-ML	10	0.83	0.79	[38]
16	Diverse set (TA98)	211	FS-ANN	6	0.86	–	[38]

[a] FS – forward selection; n/a – descriptor selection method is not specified by the author; C – clustering.

correlation coefficient R^2 = 0.77 (Table 1: #7). The model included nine different electro-topological state indexes.

Basak et al. have performed a series of studies for the prediction of mutagenicity of 95 aromatic and heteroaromatic amines. These are summarized in a recent book chapter [5]. Basak et al. have systematically explored the applicability of topological and geometric descriptors in the prediction of mutagenicity. Gradually they have extended the set of descriptors with other families of descriptors. In one of the recent studies [6] they have performed the hierarchical QSAR approach, incorporating quantum chemical descriptors, where the variables into the QSAR where selected with the modules provided by the statistical package SAS [47]. The final model included nine descriptors, with the squared correlation coefficient R^2 = 0.79 (Table 1: #8). They have further extended the database of descriptors with electro-topological state indexes [4]. The variable selection was used to reduce the number of descriptors, and a new QSAR model was obtained with the squared correlation coefficient R^2 = 0.82 (Table 1: #9).

Mazzatorta et al. have developed an artificial neural variable adaptation system (ANVAS) via combing GA and ANN methods for the selection of variables (descriptors) to QSAR models [39]. They applied it to the QSAR modeling of mutagenicity of 95 aromatic and heteroaromatic amines. The ANVAS selected 26 molecular descriptors and obtained a model with the squared correlation coefficient of R^2 = 0.64 (Table 1: #10). The selected number of descriptors is high in comparison with other models (Table 1) indicating the need to improve the selection criteria of variables in the proposed method.

We have used forward selection methods to select relevant descriptor scales for the prediction of mutagenicities on the data set of 95 aromatic and heteroaromatic amines, measured with the Ames test [37]. The final quantitative structure-activity relationship (Table 1: #11) with $R^2 = 0.83$ consisted of six descriptors related to the hydrogen bonding ability, effects induced by the surrounding medium, and the size of the compound. The model was further improved with the implicit account for nonlinear effects using the forward selection of descriptor scales and the back-propagation ANNs together [30]. The best six-parameter ANN model (6x5x1 architecture) for the mutagenicity has the squared correlation coefficient $R^2 = 0.90$ (Table 1: #12).

Recently, the work on the prediction of mutagenicities [37, 30] was extended to more diverse non-congeneric sets of compounds [38]. From literature data the diverse sets of mutagenic compounds were collected for two bacterial strains TA100 and TA98, with 177 and 212 compounds, respectively. The ML and non-linear QSAR models (Table 1: #13 - #16) were derived by employing the forward selection of descriptors from a large initial pool of theoretical molecular descriptors (in total 620) for both sets.

For the ML QSAR analysis the *best multi-linear regression* procedure was used to select the best two-parameter regression model, the best three-parameter regression model, and so on, based on the highest squared correlation coefficient, R^2, value in each step of the forward selection procedure [14]. The correlation equations were built from a pre-selected set of non-collinear descriptors. The final result had the best representation of the activity in the given descriptor pool. For the ANN procedure the experimental data (TA100 and TA98) and calculated descriptor values were scaled. Feed-forward multi-layer neural networks with input, hidden and output layers were used to represent non-linear QSAR models. A back propagation algorithm with momentum term was used to train the ANN models. The validation set error was monitored to perform automatic early stopping in order to avoid over-training of the neural network. The descriptor subsets selected with the best multi-linear regression procedure were used to adjust the initial ANN architecture for descriptor selection. The architecture with an equal number of input and hidden units was selected for both TA100 and TA98.

The significant descriptors for the ANN models were selected by a forward selection algorithm. The descriptor selection started with the pre-selection of molecular descriptors. If two descriptors were very highly inter-correlated then only one descriptor was selected. In addition, descriptors with insignificant variance were rejected. The descriptor selection algorithm started by evaluating ANN models (1x1x1) with one descriptor as input. The best models were then selected to the next step, where a new descriptor was added to the input layer and the number of hidden units was increased by one. Again, the best models were selected and this stepwise procedure was repeated until addition of new input parameters did not improve the model significantly. Since ANN models are quite likely to converge to some local minima, each model was retrained 30 times, and a model with the lowest error was selected. To speed up the training process, a larger learning rate was used and the number of allowed iterations was limited in the descriptor selection stage, since the training involves a significant amount of computations. The best models with smallest errors were selected from this procedure and were further optimized.

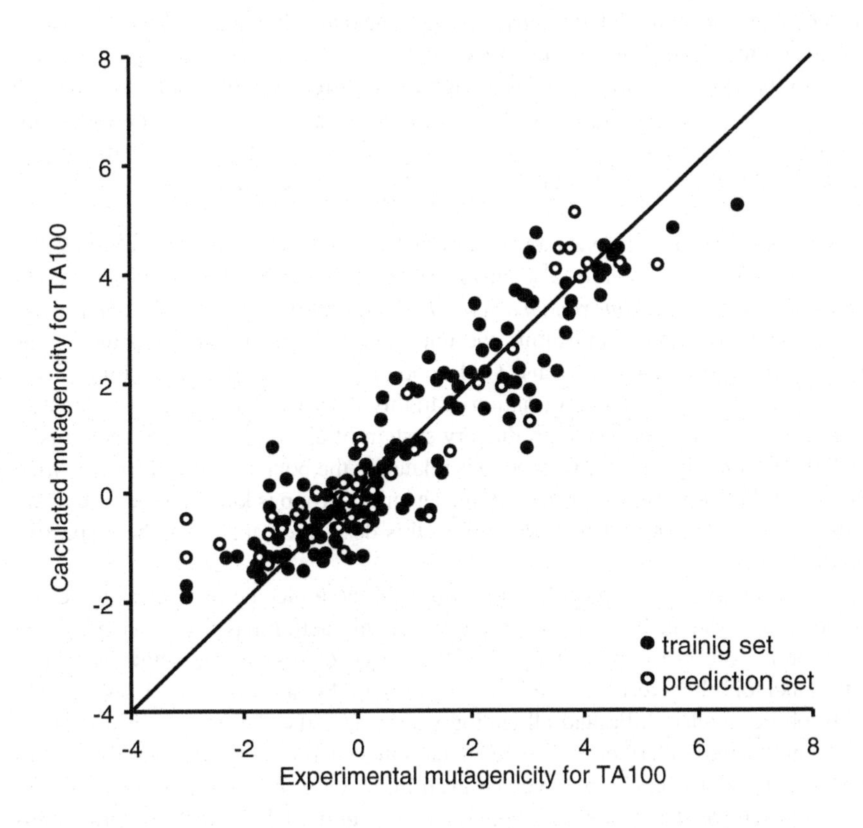

Fig. 3. Calculated vs. experimental mutagenicities derived with ANN for TA100

The stepwise development of the ML QSAR model resulted in twelve- parameter and ten-parameter equations for the TA100 and TA98 bacterial strains, respectively (Table 1: #13 & #15) [38]. The ANN development of QSAR models resulted in the seven-parameter (7x7x1 architecture) and six-parameter (6x6x1 architecture) model for the TA100 (see summarizing plot, Fig. 3) and TA98, respectively (Table 1: #14 & #16) [38]. The analysis of the descriptor content of the QSAR models shows that the size of a compound is an important determinant of mutagenicity. It can influence the transportation process of mutagenic compounds to the active site, particularly the penetration through the bio-membranes. This conclusion is also supported by our earlier work [37] where we showed that the number of rings, which is related to the size of the compounds, is a major determinant of mutagenicity for aromatic and heteroaromatic amines. Of course, the number of the rings is approximately proportional to the area of hydrophobic aromatic hydrocarbon part of the molecule and can therefore be related to the hydrophobicity of the (poly) cyclic compounds as well. Hatch and Colvin [24] have also found that the mutagenicity of aromatic and heterocyclic amines depends mainly on the size of the aromatic ring system. Both studies conclude that the size of the ring system can affect the mutagenicity in various steps during the interactions in living organisms, but most probably it affects the penetration through the bio-membranes.

Also the charge-related descriptors appear in different forms and constitute the largest group of descriptors in the derived models [38]. This clearly shows the importance of electronic effects on mutagenicity. Those descriptors cover the effects related to the specific interactions at the reactive site of the DNA and to the hydrogen bonding.

Several descriptors in our models describe the energy distribution in molecules [38]. Those descriptors can be also attributed to the reactivity of the mutagenic compounds. For instance the *maximum electrophilic reactivity index for a C atom* is common in both ML models indicating the importance of electrophilic reactivity. The presence of carbon atom, C-C and C-H bonds in the reactivity descriptors of ML models may be related to the formation of highly reactive radical centres in the aromatic systems that affect the reproductory system of a cell [44]. In the ANN model for TA98 the energy term for C-N bond is related to the bond between the nitro group and one of the carbons in the ring system. The nitro-group is known to be significant in the mutagenic action [13] and therefore this descriptor quantifies the respective reactivity.

Mutagenicity is often related to the ability of the molecule to react with the cell reproductory system. In some cases, the mutagenic action has been ascribed to an electrophilic site in the molecule [41]. In other cases, mutagenic action is related to the nucleophilic site in the molecule or to the alkylation process. However, the electrophilic, nucleophilic and alkylation sites may either be present in the parent mutagenic compound, or created as its metabolite. Therefore, several possible mechanisms of toxic action can be involved even in the case of a small group of similar compounds. Considering this, it is significant that our models show the general trends in the diverse set of compounds and bring up possible structural factors that may be important in mutagenic interactions.

The mechanism of mutagenic action includes: (1) preliminary rearrangements in the environment, (2) penetration through the biological membranes, (3) interaction with the environment in the cell (possible bio-activation to more mutagenic species) and (4) the final interaction with the mutagenic site. Proceeding from those mechanistic concepts, the interactions related to the mechanism of mutagenic action can be divided into two groups. First, the non-specific interactions that usually determine the solubility of a compound in the cell environment, penetration through the cell membranes, and hydrophobicity. The second group involves specific interactions with certain sites of the action, such as (i) electrophilic, (ii) nucleophilic and (iii) alkylation reactions. The possible metabolization of mutagenic compounds to the more reactive species may also play an important role in the specific interactions of mutagenicity. As discussed in the original articles the different descriptors can thus be related to the distinct physical interactions involved in the mechanism of mutagenic action. The non-specific interactions are modeled by the bulk structural characteristics of the compounds like the size, the surface area, the polarity, etc. The specific interactions are modeled by the electronic and energetic characteristics of the compounds described by the charge distribution or the energetic levels of the molecules.

The developed QSAR models have to be validated. This can be performed with several statistical approaches (internal cross-validation, etc.), but higher confidence is achieved when a prediction set is used, that includes compounds not used during the development of the QSAR model(s). The ML model for TA100 (Table 1: #13) predicted the mutagenicity of the external prediction set with R^2 = 0.85. The ML model for the TA98 (Table1: #15) gave the external prediction quality of R^2 = 0.82. The predictive ability of the ANN models was tested on the same prediction set. The ANN model for TA100 (Table 1: #14) predicted mutagenicity with R^2 = 0.84. The model for the TA98 (Table 1: #16) gave external prediction with R^2 = 0.86. The validation shows good prediction quality.

The comparison of ML and ANN models revealed that for both data sets, the ANN over performed the ML regression by employing much less descriptors in the model, without loss in the quality of prediction (Table 1: #13–#16). In case of TA100 the ML QSAR model is with twelve descriptors and the respective ANN QSAR model with seven descriptors. For the TA98 the ML and ANN QSAR models have ten and six descriptors, respectively. This suggests that the relationships between the structure and mutagenicity can be, in principle, nonlinear.

6 Conclusion

The use of AI techniques is gaining more and more attention in the QSAR analysis and in the connected fields. The attention is driven particularly by two major reasons. Firstly, the number of chemicals in use is growing rapidly, accompanying the myriad of information required in the research and development (e.g., drug design, design and analysis of virtual combinatorial libraries of chemicals, chemical technology implementations, etc.) and by the legislative and controlling bodies (for risk assessment and setting up production limits). Secondly, the volume of information available about chemical structures is much bigger than it was for a couple of decades. It is particularly the result of very extensive development of (theoretical) molecular descriptors, which capture either the characteristics that describe the entire molecule and/or some specific (active) site in the molecule.

The number of compounds in use can scale from several hundreds to several millions. The number of compounds in virtual combinatorial libraries scale up to several billions and the number of molecular descriptors per compound can be easily up to one thousand (it is claimed that this number can be even higher). Although the human expertise is invaluable, the amount of information we need to know and analyze is growing rapidly. Therefore computational and AI methods and tools have been actively investigated to get explanations for many chemical phenomena related to drug design, technology development and to risk assessment for better living standards.

Within the this chapter we have addressed the current state in the prediction of mutagenic potency for non-congeneric data sets in the framework of the QSAR modeling with the aid from data mining tools for the QSAR model construction. The overview of the scientific studies in the prediction of mutagenicity is given and finally a comparative study between the MLR and ANN models for TA100 and TA98 bacterial strains on non-congeneric data sets is discussed in more detail.

The relationships between mutagenicity and molecular structures are usually modeled with ML relationships. The results summarized in this chapter provide evidence that the ANNs give high quality QSAR models for the modeling of mutagenicicty. This indicates the possibility of non-linear relationship between mutagenicity and chemical structure.

As already mentioned, the experimental evaluation of mutagenicity is a difficult task. In this situation the ANN models have the advantage over ML models because they can cope with noisy data. ML regression models have an advantage because of the simple mathematical representation and are therefore much easier to interpret in terms of mechanistic concepts of mutagenicity. The extraction of rules from the ANN models, how structure determines activity, is still an open topic and a real challenge for future.

Acknowledgments

The authors are grateful to the Estonian Science Foundation (grants No. 5805 and 5386) for the financial support.

References

1. S. Agatonovic-Kustrin, R. Beresford, and A.P.M. Yusof. Theoretically derived molecular descriptors important in human intestinal absorption. *J Pharm Biomed Anal*, 25:227–237, 2001.
2. S. Agatonovic-Kustrin, L.H. Ling, S.Y. Tham, and R.G. Alany. Molecular descriptors that influence the amount of drugs transfer into human breast milk. *J Pharm Biomed Anal*, 29:103–119, 2002.
3. B.N. Ames. Identifying environmental chemicals causing mutations and cancer. *Science*, 204:587–593, 1979.
4. S.C. Basak and D. Mills. Prediction of mutagenicity utilizing a hierarchical qsar approach. *Sar Qsar Environ Rer*, 12:481–496, 2001.
5. S.C. Basak, D. Mills, B.D. Gute, and D.M. Hawkins. Predicting mutagenicity of congeneric and diverse sets of chemicals using computed molecular descriptors: A hierarchical approach. In R. Benigni, editor, *Quantitative Structure-Activity Relationship (QSAR) Models of Mutagens and Carcinogens*, pages 207–234, Boca Raton, FL, 2003. CRC Press.
6. S.C. Basak, D.R. Mills, A.T. Balaban, and B.D. Gute. Prediction of mutagenicity of aromatic and heteroaromatic amines from structure: A hierarchical qsar approach. *J Chem Inf Comp Sci*, 41:671–678, 2001.
7. R. Benigni and A. Giuliani. Mathematical models for exploring different aspects of genotoxicity and carcinogenic databases. *Environ Health Perspect*, 96:81–84, 1991.

8. R. Benigni and A. Giuliani. Quantitative structure-activity relationship (qsar) studies of mutagens and carcinogens. *Med Res Rev*, 16:267–284, 1996.
9. R. Benigni, A. Giuliani, R. Franke, and A. Gruska. Quantitative structure-activity relationships of mutagenic and carcinogenic aromatic amines. *Chem Rev*, 100:3697–3714, 2000.
10. R. Benigni and A.M. Richard. Quantitative structure-based modelling applied to characterization and prediction of chemical toxicity. *METHODS*, 14:264–276, 1998.
11. N. Bodor, A. Harget, and M.J. Huang. Neural network studies. 1. estimation of the aqueous solubility of organic compounds. *J Am Chem Soc*, 28:849–857, 1991.
12. G.G. Cash. Prediction of the genotoxicity of aromatic and heteroaromatic amines using electrotopological state indices. *Mutat Res-Gen Tox Env Mutagen*, 491:31–37, 2001.
13. A.K. Debnath, R.L.L. Compadre, D. Debnath, A.J. Shusterman, and C. Hansch. Structure-activity relationship of mutagenic aromatic and heteroaromatic nitro compounds. correlation with molecular orbital energies and hydrophobicity. *J Med Chem*, 34:786–797, 1991.
14. N.R. Draper and H. Smith. *Applied Regression Analysis*. Wiley, New York, 1981.
15. K. Enslein, V.K. Gombar, and B.W. Blake. Use of sar in computer-assisted prediction of carcinogenicity and mutagenicity of chemicals by the topkat program. *Mut Res*, 305:47–61, 1994.
16. H. Gao, M.S. Lajiness, and J. Van Drie. Enhancement of binary qsar analysis by a ga-based variable selection method. *J Mol Graph Mod*, 20:259–268, 2002.
17. A. Garg, K.L. Bhat, and C.W. Bock. Mutagenicity of aminoazobenzene dyes and related structures: a qsar/qsar investigation. *Dyes Pigments*, 55:35–52, 2002.
18. G. Gini, M. Lorenzini, E. Benfenati, P. Grasso, and M. Bruschi. Predictive carcinogenicity: A model for aromatic compounds, with nitrogen-containing substituents, based on molecular descriptors using an artificial neural network. *J Chem Inf Comput Sci*, 39:1076–1080, 1999.
19. D.E. Goldberg. *Genetic Algorithm in Search, Optimization, and Machine Learning*. Addison-Wesley, Reading, MA, 1989.
20. E.S. Goll and P.C. Jurs. Prediction of the normal boiling points of organic compounds from molecular structures with a computational neural network model. *J Chem Inf Comp Sci*, 39:974–983, 1999.
21. P. Gramatica, V. Consonni, and M. Pavan. Prediction of aromatic amines mutagenicity from theoretical molecular descriptors. *Sar Qsar Environ Res*, 14:237–250, 2003.
22. C. Hansch and T. Fujita. ρ-σ-π analysis. a method for the correlation of biological activity and chemical structure. *J Am Chem Soc*, 86:1616–1626, 1964.
23. C. Hansch and A. Leo. *Exploring QSAR - Fundamentals and Applications in Chemistry and Biology*. ACS, Washington, DC, 1995.
24. F.T. Hatch and M.E. Colvin. Quantitative structure-activity relationships (qsar) of mutagenic aromatic and heterocyclic amines. *Mut Res*, 376:87–96, 1997.
25. E. Hodgson, R.B. Mailman, and J.E. Chambers. *Dictionary of Toxicology*. Macmillan Reference Ltd, London, 1998.
26. J.J. Huuskonen, D.J. Livingstone, and A.E.P. Villa. Neural network modeling for estimation of partition coefficients based on atom-type electrotopological state indices. *J Chem Inf Comp Sci*, 38:947–955, 2000.
27. Soderman JV. *CRC Handbook of Identified Carcinogens and Noncarcinogens: Carcinogenicity and Mutagenicity Database*. CRC Press, Boca Raton, FL, 1982.
28. M. Karelson. *Molecular Descriptors in QSAR/QSPR*. John Wiley & Sons, New York, 2000.

29. M. Karelson, V.S. Lobanov, and A.R. Katritzky. Quantum-chemical descriptors in qsar/qspr studies. *Chem Rev*, 96:1027–1043, 1996.
30. M. Karelson, S. Sild, and U. Maran. Non-linear qsar treatment of genotoxicity. *Mol Simulat*, 24:229–242, 2000.
31. A.R. Katritzky, D.C. Fara, R. Petrukhin, D.B. Tatham, and U. Maran et al. The present utility and future potential for medicinal chemistry of qsar/qspr with whole molecule descriptors. *Curr Top Med Chem*, 2:1333–1356, 2002.
32. A.R. Katritzky, U. Maran, V.S. Lobanov, and M. Karelson. Structurally diverse qspr correlations of technologically relevant physical properties. *J Chem Inf Comput Sci*, 40:1–18, 2000.
33. C.D. Klaassen. *Casarett and Doull's toxicology: the basic science of poisons*. McGraw-Hill, New York, 5th edition, 1995.
34. G. Klopman and H.S. Rosenkranz. Approaches to sar in carcinogenesis and mutagenesis. prediction of carcinogenicicty/mutagenicity using multi-case. *Mut Res*, 305:33–46, 1994.
35. V.V. Kovalishyn, I.V. Tetko, A.I. Luik, V.V. Kholodovych, and A.E.P. Villa et al. Neural network studies. 3. variable selection in the cascade-correlation learning architecture. *J Chem Inf Comp Sci*, 38:651–659, 1998.
36. J.P. Lowe. *Quantum Chemistry*. Academic Press, San Diego, 1993.
37. U. Maran, M. Karleson, and A.R. Katritzky. A comprehensive qsar treatment of the genotoxicity of heteroaromatic amines. *Quant Struct-Act Relat*, 18:3–10, 1999.
38. U. Maran and S. Sild. Qsar modeling of genotoxicity on non-congeneric sets of organic compounds. *Artif Intell Rev*, 20:13–38, 2003.
39. P. Mazzatorta, M. Vračko, and E. Benfenati. Anvas: Artificial neural variables adaptation system for descriptor selection. *J Comput Aid Mol Des*, 17:335–346, 2003.
40. J. McCann and B.N. Ames. The salmonella/microsome mutagenicity test: Predictive value of animal carcinogenicity. In H.H. Hiatt, J.D. Watson, and J.A. Winsten, editors, *Origins of Human Cancer: Human Risk Assessment*, volume C, pages 1431–1450. Cold Spring Harbor Laboratory, Cold Spring Harbor, NY, 1977.
41. J.A. Miller and E.C. Miller. Ultimate chemical carcinogens as reactive mutagenic electrophiles. In H.H. Hiatt, J.D. Watson, and J.A. Winsten, editors, *Origins of Human Cancer*, volume B, pages 605–627. Cold Spring Harbor Laboratory, Cold Spring Harbor, NY, 1977.
42. B.E. Mitchell and P.C. Jurs. Prediction of autoignition temperatures of organic compounds from molecular structure. *J Chem Inf Comput Sci*, 37:538–547, 1997.
43. S.J. Patankar and P.C. Jurs. Prediction of ic50 values for acat inhibitors from molecular structure. *J Chem Inf Comp Sci*, 40:706–723, 2000.
44. A. Pullman and B. Pullman. Electronic structure and carcinogenic cctivity of some aromatic molecules. *Adv Cancer Res*, 3:117–169, 1955.
45. A.M. Richard. Application of sar methods to non-congeneric data bases associated with carcinogenicity and mutagenicity: Issues and approaches. *Mut Res*, 305:73–97, 1994.
46. D. Rogers and A.J. Hopfinger. Application of genetic function approximation to quantitative structure-activity relationships and quantitative structure-property relationships. *J Chem Inf Comp Sci*, 34:854–866, 1994.
47. http://www.sas.com/.
48. T.W. Sawyer. Cellular methods of genotoxicity and carcinogenicity. In F.A. Barile, editor, *In Vitro Cytotoxicology: Mechanisms and Methods*, pages 127–174, Boca Raton, FL, 1994. CRC Press.
49. S. Sild and M. Karelson. A general qspr treatment for dielectric constants of organic compounds. *J Chem Inf Comp Sci*, 42:360–367, 2002.

50. S.S. So, S.P. van Helden, V.J. van Geerestein, and M. Karplus. Quantitative structure-activity relationship studies of progesterone receptor binding steroids. *J Chem Inf Comp Sci*, 40:762–772, 2000.
51. P. Somol, J. Pudil, J. Novovičová, and P. Paclík. Adaptive floating search methods in feature selection. *Pat Rec Lett*, 20:1157–1163, 1999.
52. L. Sztandera, A. Garg, S. Hayik, K.L. Bhat, and C.W. Bock. Mutagenicity of aminoazo dyes and their reductive-cleavage metabolites: a qsar/qsar investigation. *Dyes Pigments*, 59:117–133, 2003.
53. I.V. Tetko, A.E.P. Villa, and D.J. Livingstone. Neural network studies. 2. variable selection. *J Chem Inf Comp Sci*, 36:794–803, 1996.
54. R. Todeschini and V. Consonni. A new algorithm for optimal, distance-based experimental design. *Chemom Intell Lab Syst*, 16:37–44, 1992.
55. R. Todeschini and V. Consonni. *Handbook of Molecular desciptors*. Wiley-VCH, Weinheim, 2000.
56. V.N. Viswanadhan, G.A. Mueller, S.C. Basak, and J.N. Weinstein. Comparison of a neural net-based qsar algorithm (pcann) with hologram- and multiple linear regression-based qsar approaches: Application to 1,4-dihydropyridine based calcium channel antagonists. *J Chem Inf Comp Sci*, 41:505–511, 2001.
57. B. Walczak and W. Wegscheider. Non-linear modeling of chemical data by combinations of linear and neural net methods. *Anal Chim Acta*, 283:508–517, 1993.

Characterizing Gene Expression Time Series using a Hidden Markov Model

Sally McClean, Bryan Scotney and Steve Robinson

School of Computing and Information Engineering, Faculty of Informatics,
University of Ulster, Northern Ireland.
E-mail: {si.mcclean, bw.scotney,s.robinson}@ulster.ac.uk

Summary. We are concerned with the temporal clustering of a series of gene expression data using a *hidden Markov model* (HMM) and in so doing providing an intuitive way of characterizing the developmental processes within the cell. By explicitly modelling the time dependent aspects of these data using a novel form of the HMM, each stage of cell development can be depicted. In this model, the hitherto unknown development process that manifests itself as changes in gene expression is represented by hidden concepts. We use clustering to learn probabilistic descriptions of these hidden concepts in terms of a *hidden Markov process*. Finally, we derive linguistic identifiers from the transition matrices that characterize the developmental processes. Such identifiers could be used to annotate a genome database to assist data retrieval.

1 Introduction

Microarray technology has proved to be a useful tool for measuring the expression level of thousands of genes in a single experiment. Using a series of experiments enables the gathering of large quantities of gene expression data under various experimental conditions or at different times. However, the biological interpretation of these gene expression data is left to the observer and due to its sheer volume and the complexity of possible ways that genes might be structurally and functionally related, presents a formidable challenge. Current methods for the analysis of gene expression data typically rely on algorithms that cluster genes with respect to a series of states [11] or use other computational tools to organize the gene expression profiles into conceptual schemes. The biological premise underlying the use of clustering algorithms is that genes that display similar temporal expression patterns are co-regulated and may share a common function or contribute to common pathways in the cellular processes. These clustering techniques have relied on a variety of association metrics such as Euclidean distance and correlation coefficients. However, such an approach can only achieve a limited understanding of dynamic processes as has been pointed out by Cadez et al. [2].

In this paper, our aim is to provide a model that will help to understand the semantics of the underlying processes within the cell in much the same way as a linguist may analyze a hitherto unknown script. For the time series aspects of the gene

W. Dubitzky and F. Azuaje (eds.), Artificial Intelligence Methods and Tools for Systems Biology, 37–50.

expression, we adopt a *hidden Markov model* (HMM), where the states represent the stages of the underlying process concept. By concept we mean a notion of interest, for example, we may define adolescent as a person aged between 12 and 16 years. Concept learning is an important task in machine learning [12]. By process we mean a dynamic behavior, such as growth, that typically proceeds through a number of stages. Here, the concept being studied is the central nervous system of a rat that experiences the process of development. For example, we may associate a gene cluster with an underlying growth process that develops through a number of stages that are paralleled by developments in the associated genes.

Each cluster is represented by a concept, which is described in terms of a HMM. We seek to find schema mappings between the states of the observed expression time series in the data and the states of the HMM (the model). There are several advantages to this approach. It explicitly models the temporal aspects of the data in terms of stages, thereby providing an intuitive way of describing the underlying behavior of the process. It also allows us to use probabilistic semantics, inherent in the model, to take account of natural variability in the process. Such segmental models have considerable potential for time sequence data [6]. The characterization of the clusters in terms of a hidden underlying process concept may correspond to a hitherto unknown biological process that may only later be recognized and understood.

This approach is extended to the integration and clustering of non-homogeneous Markov chains to reflect the temporal semantics of the data. A similar method for homogeneous data is described in Smyth [14] and Cadez et al. [2]. We have previously assumed that the schema mappings are available *ab initio.*

In this paper the novelty resides mainly in the fact that we must now learn the schema mappings as well as the cluster memberships and profiles. Our use of non-time-homogeneous HMMs in this context is also novel. Such constructs allow us to better describe the temporal semantics of a dynamic process in a stochastic environment.

Model-based clustering is well established and, in general, provides a flexible methodology that can cope with complex and heterogeneous data. Clustering is based on the measurement of similarity between the data and the model. The alternative approach, using distance metrics, tends to be less robust in terms of its ability to cope with complicated data that may be statistically incomplete in various regards [2, 3]. Our approach is intermediate between these two methodologies in that the initial gene expressions are clustered using a distance metric based on mutual information, that is, a similar temporal profile. A full model-based approach, such as that described by Cadez for homogeneous schema, could also be utilized here; the advantage of our strategy is that we decouple clustering from schema learning, thus allowing simpler ad hoc methods to be substituted for these tasks in the interests of simplicity and computational efficiency.

2 Overview

2.1 The general process

We divide the problem into a number of distinct tasks. For each set of gene expression time series data within the data set, we identify genes that have a similar temporal signature. These genes are then grouped to form gene clusters. Although the genes within a cluster exhibit comparable patterns of expression, the time indexing of the data themselves may be heterogeneous. The term heterogeneous here is used in the sense that they may represent either different attributes, the same attribute held at different granularities or indeed the series may be of differing length. This is often the case where the data is from diverse origins. Such issues are resolved by mapping each of the expression data sets to an idealized though previously unknown time series data set that may be deemed to be a representative expression of the underlying hidden process concept. This data set must be learned, and once learned, we refer to it as the *hidden process*. On the basis of these mappings we may learn the probabilistic description of this process.

Where an underlying and maybe unknown dynamic process is driving an observed set of events, the hidden Markov model is a valuable mathematical tool. The HMM describes a latent process, usually associated with function, that changes state in a similar way to the observed gene expression. In some respects, this latent process may be considered to be driving the observed events.

The model defines a set of states, $h_1, \cdots, h_k$ each of which is associated with a multi-dimensional probability distribution called a transition matrix $\mathbf{P} = \{p_{ij}\}$. Transitions between the states are governed by these probabilities called transition probabilities where:

$$p_{ij} = \text{Prob}\ \{\text{HMM is at } h_j \text{ at time } t+1 \text{ given the HMM is at } h_i \text{ at } t\}.$$

From an intial state defined by the *initial state* vector $\mathbf{s}_0 = \{s_{0j}\}$ any particular outcome can be generated, according to the associated transition matrix. However, in general, only the outcome and not the state itself is observable, hence the name hidden Markov model. Fig. 1 illustrates the relationship between the state space in the HMM and the elements of the state vector.

Working with a cluster of genes that exhibit a similar temporal expression pattern, our goal is to determine a temporal clustering of the time series. Each temporal cluster is then characterized by a HMM transition matrix, the symbols of each series in the cluster can then be mapped onto states of the HMM.

When the temporal processes within the gene cluster are described in terms of this HMM, the states of the model can be said to characterize the behavior of that gene cluster. We propose a system of formal linguistic identifiers, based on the general form of the HMM states, that would represent each class of behavior. If a gene expression database were annotated with such identifiers, this would facilitate the search and retrieval of gene clusters with a given behavior.

While, in general terms, these tasks may be carried out in an integrated manner, as an initial strategy we have decided to approach each of these tasks independently.

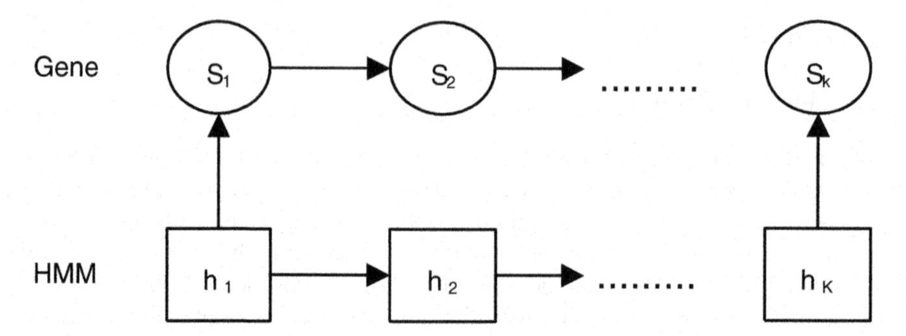

Fig. 1. HMM state space and state vector elements.

We recognize two distinct sub-problems. The first sub-problem involves mapping and re-labelling of the respective time series alphabets. In order to do this we need to recognize the sequences where re-labelling is appropriate and then learn the optimal labelling. The second sub-problem involves learning a HMM based on these mapped time series. Then, by using this HMM to identify and characterize the temporal clusters, those clusters can be attributed with an appropriate linguistic label.

2.2 An illustrative example

We consider the problem of clustering heterogeneous time series of symbolic data. The series have in common the fact that, within a cluster, they are related to a common underlying concept process that is conceptualized here as a hidden Markov model. As an example we present four time series of gene expression adapted from D'haeseleer et al. [4]. Here the letters A, B, etc. may, for example, be discretized numerical ranges. In Table 1 the four series, {Series 1, Series 2} and {Series 3, Series 4} form two separate clusters when we perform the mappings presented in Table 2.

Table 1. Gene expression data

Series 1	A	A	B	B	B	B	B	B	B
Series 2	C	D	E	E	E	E	E	E	E
Series 3	F	F	F	G	G	G	G	H	H
Series 4	I	I	I	K	K	K	K	J	J

Table 2. Schema mappings for Table 1

Cluster 1 $S^{1\&2}$	U	U/V	W	W	W	W	W	W	W
Cluster 2 $S^{3\&4}$	X	X	X	Y	Y	Y	Y	Z	Z

The series (U, U/V, W, W, W, W, W, W, W) and (X, X, X, Y, Y, Y, Y, Z, Z) therefore characterize the behavior of the HMMs for the latent variables underpinning clusters 1 and 2 respectively. Here U/V means that this term may be either U or V. Although, in some circumstances, the domain expert may know such schema mappings, typically they are unknown and must be discovered by the algorithm.

We define a correspondence table for a set of series to be a table that represents mappings between the schemes. The correspondence table for cluster 1 (series 1 and 2) is shown in Table 3; the symbolic values in each series are described alphabetically.

Table 3. The correspondence table for cluster 1 in Table 1.

Hidden Concept 1	Series 1	Series 2
U	A	C
V	-	D
W	B	E

This correspondence table must be learned in order to determine the mappings between heterogeneous time series.

Earlier work has used mutual entropy to cluster data of this type [2]. Examination of the mutual entropy identifies where the symbols in the respective sequence alphabets co-occur. Finding these schema mappings would involve searching over all the possible mappings. However, another possibility is to use a heuristic approach such as a genetic algorithm to minimize the divergence between the mapped time series. We may limit the types of mapping that are permissible in order to restrict the search space e.g., we may allow only order-preserving mappings. The fitness function may also be penalized to prohibit trivial mappings, e.g., where every symbol in the series is mapped onto the same symbol of the HMM.

While the use of mutual entropy has been criticized previously on the grounds that it does not take the temporal aspects into account, we nonetheless adopt it in our initial approach, though only to determine mappings between the heterogeneous time series in order to identify homogeneous clusters. When testing whether to combine two clusters, we calculate the mutual information for the two clusters separately and then the two clusters combined. If the value of H_{12} (the pooled entropy) is inside a threshold of $H_1 + H_2$ (the sum of the separate entropies), then clusters 1 and 2 are sufficiently close together to be combined. This metric is chosen because it tends to cluster together sequences whose values are associated, i.e., they tend to co-occur. This means that within a cluster we can find a transformation such that mapping associated values onto a common ontology leads to sequences that are similar.

2.3 Learning the mapping

The schema mappings from each of the gene expression time series to the hidden process ensure that the symbols of the series are of the same alphabet. These mappings may also serve another purpose as mentioned above, of facilitating the difference in granularity of the time series to the hidden process. Although it is possible in theory for the data to be of coarser granularity than the hidden process, we restrict our attention to mappings to a hidden process that may be coarser. We make this assumption partly because the alternative would raise methodological difficulties such as problems of non-identifiability.

Our objective is to carry out the mapping such as to minimize the distance between each gene expression series and the hidden process. Since the search space for optimal mappings is potentially very large and the hidden process is a priori unknown, we propose to approximate this function with an ad hoc approach that can be used to initialize a heuristic method. Our initialization method then finds a mapping by choosing one of the gene expression time series whose number of symbols is maximal and each of those symbols then acts as a proxy for the symbols of the hidden process. For each of the remaining series, each symbol is mapped to one of the symbols of the hidden process in such a way that the sum of the distances in that time series is minimal. If we wish to provide a number of solutions to form the initial mating pool for a genetic algorithm, we can choose a number of different series to act as proxies for the HMM.

2.4 Learning the HMMs

We now develop an algorithm for learning the transition matrix of the HMM for a set of heterogeneous time series, given that the mappings between the respective series schema is a priori known. This matrix then characterizes the underlying process described by the HMM.

Maximum likelihood is a common statistical method for estimating probabilities. Using this approach we find that:

$$p_{ij} = \frac{\sum_t \text{numbers of transitions from state } i \text{ to state } j \text{ at time } t}{\sum_t \text{total numbers of genes in state } i \text{ at time } t} \quad (1)$$

and

$$p_{0j} = \frac{\text{total number of genes in state } j \text{ at time } t = 0}{\text{total number of genes at time } t = 0} \quad (2)$$

where t is the development time point.

Further details of these equations may be found in McClean, et al. [10].

2.5 Characterizing the clusters

Once we have learned the HMM probabilities using equations (1) and (2), the next task is to use these probabilities to characterize the clusters in terms of the optimal pathway for each cluster. We use *dynamic programming* (DP) to compute the optimal pathway for each HMM. This is a special case of a standard technique in the theory of HMM, where such a DP approach is known as the Viterbi Algorithm [14]. The optimal pathway is here chosen so that the probability of the next step is as large as possible.

2.6 The time-inhomogeneous HMM

The HMM is typically used to model an underlying process that produces a succession of observed states. Our concept here has been to use the HMM model to provide a means of associating the expression of a gene cluster with an underlying process concept as it moves through a succession of developmental stages. While it is possible that the HMM is time invariant, in these cases, the underlying transition matrix of the HMM is a function of time changing in response to external stimuli. As we will see in the next section, processes such as we have in mind are likely to exhibit such time heterogeneous behavior.

Once the hidden process has been determined, its state at each time point is characterized by a HMM transition matrix. Now we wish to create a set of temporal clusters of contiguous time periods where each temporal cluster is subject to a time-homogeneous Markov model. Thus, changes in the underlying process can be traced to a series of change points and the prevailing situations between these time points identified.

Beginning at the first time point, contiguous matrices are tested for similarity and if they are found not to be significantly different, they are combined and compared with the following matrices. These combined transition matrices then represent a uniform contiguous series of gene expression data. If however, the next contiguous matrix fails the test of similarity, then that cluster is deemed to have ended and a new cluster commences. Here a temporal cluster may be thought of as a pattern that prevails for a period of time and then changes to a different pattern.

The similarity metric is given by $L_{12} - L_1 - L_2$, where L is the log-likelihood function for the elements of the transition matrix given the observed transitions. While the log-likelihood is a similar distance metric to the mutual entropy used earlier for gene clustering, we are now carrying out temporal clustering, by comparing transition matrices at the respective time points. The previous gene clustering was static, involving the comparison of probabilities of occurrence of the series states. We now adopt here the previous approach of McClean et al. [8, 9] that uses likelihood ratio tests in a sequential manner to cluster Markov chains at contiguous time points, where appropriate.

It is here required to characterize each of the newly formed temporal clusters by determining the optimal pathway. This procedure, as before, is achieved through dynamic programming to combine optimal sub-pathways to find the overall optimal pathway.

3 Application

The procedures outlined above are now illustrated using temporal series of gene expression data. These data are available electronically at: `http://stein.cshl.org/genome_informatics/expression/somogyi.txt` and have been analyzed in a number of papers [11, 4]. The gene activity data refers to the expression of 112 genes within the spinal chord of a rat during its development from embryo to adult. Only the specific genes that are considered to be important in the process, whereby the cells of the central nervous system become specialized nerve cells, have been analyzed. Indeed these genes are known to code for the manufacture of various proteins associated with the role of a nerve cell. These proteins include neurotransmitter receptor proteins (Acetylcholine receptors: nAChRa2, mAChR2, mAChR3, nAChRa3, nAChRa7), peptide signalling proteins (Heparin-binding growth factors: MK2, EGFR, PDGFR) and one neuroglial marker (NFL). The data contain a time series of gene expression measured at eight developmental time points E11, E15, E18, E21, P0, P7, P14, A, referring to *embryonic*, *postnatal*, and *adult* phases respectively. The scales of gene expression have been discretized by partitioning the expressions into three equally sized bins and domain experts have identified the gene clusters based on this three-bin discretization. In this analysis the discretization has a smoothing effect that facilitates characterizations of the time series without unduly masking the underlying pattern. Associations between the gene expression time series were identified using mutual entropy and the clusters based on this distance metric are described in detail in Michaels et al. [11].

Of the clusters identified, we now use cluster 3, the smallest cluster, to illustrate our approach. Using this cluster, we first learn the mapping that characterizes the cluster. Once we have succeeded in mapping the sequences to a hidden process, we can characterize that hidden process in terms of the optimal pathway using the initial vector and transition matrix of the underlying Markov chain. In this case the gene expression time series are presented in Table 4 where the codes 0, 1, and 2 are local references and therefore have different meanings in each gene.

A heuristic process was then used to learn the mappings between the data in Table 4 and the hidden process. The heuristic process used a randomized search method to optimize sequence alignment by finding the most probable mapping. These mappings are shown in Table 5 and the transformed sequences are illustrated in Table 6.

Table 4. Time series of gene expression for Cluster 3.

Gene	E11	E13	E15	E18	E21	P0	P7	P14	A
nAChRa2_RATNNAR	0	0	0	1	2	2	2	2	1
mAChR2_BOVMRM2SUB	0	0	0	2	2	2	2	2	1
mAChR3_RATACHRMB	0	0	0	1	2	2	1	1	0
nAChRa3_RNACHRAR	0	0	2	1	2	2	1	1	0
EGFR_RATEGFR	0	1	0	1	2	2	2	2	0
NFL_RATNFL	0	1	1	1	1	2	2	1	0
nAChRa7_RATNARAD	1	0	0	2	2	2	2	2	1
MK2_MUSMK	2	2	2	1	1	1	1	1	0
PDGFR_RNPDGFRBE	2	2	2	0	0	0	0	0	1

Table 5. The mappings for the time series in Table 4.

Gene	Mapped Sequence Values		
	0	1	2
nAChRa2_RATNNAR	0	1	2
mAChR2_BOVMRM2SUB	0	0	2
mAChR3_RATACHRMB	0	2	2
nAChRa3_RNACHRAR	0	2	2
EGFR_RATEGFR	0	0	2
NFL_RATNFL	0	1	2
nAChRa7_RATNARAD	0	0	2
MK2_MUSMK	1	2	0
PDGFR_RNPDGFRBE	2	1	0

Table 6. The series in Table 4 mapped using the transformations in Table 5.

Gene	E11	E13	E15	E18	E21	P0	P7	P14	A
nAChRa2_RATNNAR	0	0	0	1	2	2	2	2	1
mAChR2_BOVMRM2SUB	0	0	0	2	2	2	2	2	0
mAChR3_RATACHRMB	0	0	0	2	2	2	2	2	0
nAChRa3_RNACHRAR	0	0	2	2	2	2	2	2	0
EGFR_RATEGFR	0	0	0	0	2	2	2	2	0
NFL_RATNFL	0	1	1	1	1	2	2	1	0
nAChRa7_RATNARAD	0	0	0	2	2	2	2	2	0
MK2_MUSMK	0	0	0	2	2	2	2	2	1
PDGFR_RNPDGFRBE	0	0	0	2	2	2	2	2	1

Using the transformed sequences in Table 6 the maximum likelihood estimates of the initial vector and transition matrix were then found to be:

$$\mathbf{s}_0 = \begin{pmatrix} 1 \\ 0 \\ 0 \end{pmatrix} \qquad \mathbf{P} = \begin{pmatrix} 0.64 & 0.08 & 0.28 \\ 0.17 & 0.5 & 0.33 \\ 0.12 & 0.10 & 0.78 \end{pmatrix}$$

The interpretation of this transition matrix is that all sequences start in state 0, and there is a probability 0.64 of still being there, 0.08 of being in state 1 and 0.28 of being in state 2, at the next time point. Using dynamic programming with $\mathbf{s}_0$ and $\mathbf{P}$, as described above, we learn the most probable pathway given the initial state vector and the transition matrix: this is (0, 2, 2, 2, 2, 2, 2, 2, 2). By inspection of the mapped sequences in Table 6, it appears that this pathway is not, in fact, very representative of the process. The reason for this is probably because of a problem with time heterogeneity, as described above. We therefore now apply the methodology discussed in that section for time-inhomogeneous HMMs.

Using the mapped sequence data in Table 6, the separate transition matrices are found to be:

$$\mathbf{P}_{E11} = \begin{pmatrix} 0.89 & 0.11 & 0 \\ * & * & * \\ * & * & * \end{pmatrix} \qquad \mathbf{P}_{E13} = \begin{pmatrix} 0.87 & 0 & 0.13 \\ 0 & 1 & 0 \\ * & * & * \end{pmatrix}$$

$$\mathbf{P}_{E15} = \begin{pmatrix} 0.14 & 0.14 & 0.72 \\ 0 & 1 & 0 \\ 0 & 0 & 1 \end{pmatrix} \qquad \mathbf{P}_{E18} = \begin{pmatrix} 0 & 0 & 1 \\ 0 & 0.5 & 0.5 \\ 0 & 0 & 1 \end{pmatrix}$$

$$\mathbf{P}_{E21} = \begin{pmatrix} * & * & * \\ 0 & 0 & 1 \\ 0 & 0 & 1 \end{pmatrix} \qquad \mathbf{P}_{P0} = \begin{pmatrix} * & * & * \\ * & * & * \\ 0 & 0 & 1 \end{pmatrix}$$

$$\mathbf{P}_{P7} = \begin{pmatrix} * & * & * \\ * & * & * \\ 0 & 0.11 & 0.89 \end{pmatrix} \qquad \mathbf{P}_{P14} = \begin{pmatrix} * & * & * \\ 1 & 0 & 0 \\ 0.62 & 0.38 & 0 \end{pmatrix}$$

Here $*$ denotes that there were no data available to estimate this probability. Comparing $\mathbf{P}_{E11}$ with $\mathbf{P}_{E13}$ the log-likelihood distance is 2.78; the chi-squared statistic with 2 degrees of freedom is 5.99 at a significance level of 95%, suggesting that we should decide that there is no significant difference between $\mathbf{P}_{E11}$ and $\mathbf{P}_{E13}$. These matrices are therefore combined and the combined matrix compared with $\mathbf{P}_{E15}$. This time the log-likelihood distance is 15.82 compared with a test statistic of 5.99, so we decide that E15 is in a different cluster to that containing E11 and E13. Proceeding in a similar fashion, we find that $\mathbf{P}_{E15}$ can be combined with $\mathbf{P}_{E18}$, $\mathbf{P}_{E21}$, $\mathbf{P}_{P0}$, and $\mathbf{P}_{P7}$. However, when we compare the combined transition matrix from $\{\mathbf{P}_{E15}, \mathbf{P}_{E18}, \mathbf{P}_{E21}, \mathbf{P}_{P0}, \text{and } \mathbf{P}_{P7}\}$ with $\mathbf{P}_{P14}$ the log- likelihood distance is 40.98; the chi-squared statistic with 4 degrees of freedom is 9.49 at a significance level of 95%, suggesting that we should decide that there is a significant difference. There are therefore three temporal clusters: {E11, E13}, {E15, E18, E21, P0, P7, P14}, and {A}, with respective transition matrices:

$$\mathbf{P}_1 = \begin{pmatrix} 0.88 & 0.06 & 0.06 \\ 0 & 1 & 0 \\ * & * & * \end{pmatrix} \quad \mathbf{P}_2 = \begin{pmatrix} 0.125 & 0.125 & 0.75 \\ 0 & 0.5 & 0.5 \\ 0 & 0.03 & 0.97 \end{pmatrix} \quad \mathbf{P}_3 = \begin{pmatrix} * & * & * \\ 1 & 0 & 0 \\ 0.62 & 0.38 & 0 \end{pmatrix}$$

We now use this time-inhomogeneous HMM to re-compute the optimal pathway. This is given by: (0, 0, 2, 2, 2, 2, 2, 2, 0). We see from Table 6 that this is much more plausible than the optimal pathway predicted by the time homogeneous HMM; in fact three of the observed sequences coincide with it. We have therefore demonstrated that such problems may benefit from the time-inhomogeneous approach.

The model describes the group of genes showing marked changes in gene activity over a period of time from halfway into the embryonic stage of the rat (E15) until just before the adult stage (A). The marked change can be either an increase in the level of activity or a decrease in the level of activity. The peptides produced by some genes are known to lead to the promotion or inhibition of another gene later on in time. The importance of this model is that it can demonstrate which genes change in their activity over time in either direction and thereby provide an indicator of a causal relationship between the genes. Although only a few genes have been analyzed here, a more complete study would provide a significant tool for the understanding of processes within the development of the cell.

4 Learning linguistic descriptions of the sequences

4.1 The homogeneous case

Although gene expression databases are burgeoning, tools for their exploration are in their infancy [1]. An important aim of such tools is the interpretation of gene expression time series in terms of changing states of the organism. In seeking to relate gene activity to known processes and pathways, we might match our characterization, defined by the states of the HMM, to the function of the genes. While functional data are not always available or may be incomplete, linguistic descriptions, derived from signature patterns, could be used to describe the temporal semantics of the series. Such annotations may then be used to augment gene expression databases. Although such linguistic labelling of genes has been used in the past [5, 7, 13], the annotation of temporal sequences is novel.

We have identified three terms that capture the temporal semantics of commonly occurring sequence patterns for a homogeneous HMM, namely: *constant, random,* and *trend.* We characterize these as follows:

Constant

A constant pattern is characterized by the HMM transition matrix being the identity matrix. Thus the gene stays in its initial state and never moves to an alternative state. Since the real world is usually subject to some noise, this definition is relaxed to one where the process is quasi-constant, and the matrix is 'nearly' the identity, reflecting very little changes in states. Here we can define 'nearly' in terms of thresholds e.g., all diagonal elements of the transition matrix should be greater than or equal to 0.9. An example of this pattern is therefore one that has the transition matrix:

$$\mathbf{P} = \begin{pmatrix} 0.9 & 0.05 & 0.05 \\ 0.03 & 0.92 & 0.05 \\ 0.03 & 0.04 & 0.93 \end{pmatrix}$$

Random

A Markov chain can be characterized by its limiting behavior as $t \rightarrow \infty$. Often the Markov chain can be described as ergodic, that is that the limiting probabilities of being in each state, when the transition matrix is $\mathbf{P}$, is given by the eigenvector of $\mathbf{P}$ corresponding to the eigenvalue 1. An important special case of an ergodic Markov chain is one which has every element positive. We here use the term *random* of processes which are quasi-ergodic; this may be determined by thresholding the steady state vector, e.g., if all elements are greater than 0.1. An example of a pattern that we define as *random* is therefore one that has transition matrix

$$\mathbf{P} = \begin{pmatrix} 0.5 & 0.3 & 0.2 \\ 0.4 & 0.25 & 0.35 \\ 0.2 & 0.25 & 0.55 \end{pmatrix}$$

with corresponding steady state vector given by

$$\pi = \begin{pmatrix} 0.36 \\ 0.27 \\ 0.37 \end{pmatrix}$$

This means that there is a lot of change between states but there is no particular pattern to this.

Trend

Another important type of Markov chain has states that do not communicate with each other i.e. the probability of moving between them in any number of steps is zero. In this case we say that the Markov chain is reducible into classes where each class consists of a group of states that do not communicate with other states. Where there is zero probability of exiting a class, once entered, we say that this class is persistent. We here use the term *trend* of a process which initiates in a transient state and moves to a persistent state. *Trend* is then characterized by a limiting distribution which has zeros corresponding to non-zeros in the initial vector and vice versa vector; the quasi reducible process may be determined by thresholding the initial and steady state vector, e.g., when appropriate elements are greater than 0.9. An example of a pattern that we define as *trend* is therefore one which has the transition matrix:

$$\mathbf{P} = \begin{pmatrix} 0.2 & 0.3 & 0.5 \\ 0 & 0.4 & 0.6 \\ 0 & 0.1 & 0.9 \end{pmatrix} \text{ or } \mathbf{P} = \begin{pmatrix} 0.85 & 0.15 & 0 \\ 0.7 & 0.2 & 0.1 \\ 0.6 & 0.35 & 0.05 \end{pmatrix}$$

with corresponding initial vectors given by:

$$\mathbf{s} = \begin{pmatrix} 1 \\ 0 \\ 0 \end{pmatrix} \text{ and } \pi = \begin{pmatrix} 0 \\ 0 \\ 1 \end{pmatrix}$$

and steady state vector given by

$$\pi = \begin{pmatrix} 0.82 \\ 0.16 \\ 0.02 \end{pmatrix}$$

4.2 The inhomogeneous case

In the inhomogeneous case, we use one of the labels *constant, random,* and *trend* within each temporal cluster. Thus, using the temporal clusters identified in Section 3 for the rat data, we would describe the pattern as:

Constant (at 0) in E11, E13,
Trend (to 2) in E15, E18, E21, P0, P7, P14, and
Random (in 0 and 1) inA.

In fact it is likely that the third pattern reverts to the first one, once the genes have settled into their adult behavior. However, with only one adult observation in this data set we cannot infer this at the moment. Such a pattern, which moves from one constant state to another constant state, staying in each constant state for an amount of time that is determined by the underlying process, is likely to be commonly occurring. We label this pattern as *staged,* where we envisage moving through a series of stages in response to stages in the underlying process.

5 Conclusions

We have described a novel solution to a sequence-clustering problem where the sequences have been classified according to heterogeneous classification schemes. We adopt a model-based approach that uses a hidden Markov model that has, as states, the stages of the underlying process that generates the sequences. Each cluster is described in terms of a HMM where we seek to find schema mappings between the states of the original sequences and the states of the HMM.

The general solution that we propose involves a complex process in which we first cluster the heterogeneous time series, next determine mappings between the sequences and the HMM, then find the parameters for each cluster-specific HMM, and finally describe the clusters in terms of optimal pathways. Time-inhomogeneous HMMs are used to test for time-inhomogeneity and, where appropriate, HMM transition matrices and optimal pathways are determined for the separate temporal clusters. Finally we identify linguistic identifiers to facilitate annotation of gene expression data.

This approach is used to cluster and characterize heterogeneous gene expression time series data; for illustrative purposes we use a data set that is publicly available on the Internet. Although this is a modest data set, it serves to explain our approach and demonstrates the necessity of considering time-inhomogeneity for such problems, where there is an underlying temporal process involving staged development.

In conclusion, we believe that this paper has identified, and provided a framework for solving, an important problem for large, possibly distributed, heterogeneous gene expression time series data.

Acknowledgments

This work was partially supported by the MISSION (Multi-agent Integration of Shared Statistical Information over the (inter)Net) project, IST project number 1999-10655, which is part of *Eurostat's* EPROS initiative funded by the European Commission.

References

1. D.E.Jr. Bassett, M.B. Eisen, and M.S. Boguski. Gene expression informatics - it's all in your mine. *Nature Genetics Supplement*, 21:41–55, 1999.
2. I. Cadez, S. Gaffney, and P. Smyth. A general probabilistic framework for clustering individuals. In *ACM SIGKDD*, pages 140–149, 2000.
3. I.V. Cadez, D. Heckerman, C. Meek, P. Smyth, and S. White. Model-based clustering and visualisation of navigation patterns on a web site. *Journal of Data Mining and Knowledge Discovery*, 7, 2003.
4. P. D'haeseleer, X. Wen, S. Fuhrman, and R. Somogyi. Mining the gene expression matrix: Inferring gene relationships from large scale gene expression data. In Paton RC Holcombe M, editor, *Information Processing in Cells and Tissues*. Plenum Publishing, 1998.
5. F. Eisenhaber and P. Bork. Evaluation of human readable annotation in biomolecular sequence databases with biological rule libraries. *Bioinformatics*, 15, 1999.
6. X. Ge and P. Smyth. Deformable markov model templates for time-series pattern matching. In *ACM SIGKDD*, pages 81–90, 2000.
7. D.R. Masys, J.B. Welsh, J.L. Fink, M. Gribskov, and I. Klacansky et al. Use of keyword hierarchies to interpret gene expression patterns. *Bioinformatics*, 17, 2001.
8. S. McClean, E. Montgomery, and F. Ugwouwo. Non-homogeneous continuous time markov and semi-markov manpower models. *Applied Stochastic Models and Data Analysis*, 13:191–198, 1998.
9. S.I. McClean, B.W. Scotney, and K.R.C. Greer. Clustering heterogenous distributed databases. In Kumar V Obradovic Z Kargupta H, Ghosh J, editor, *Proceedings of KDD Workshop on Knowledge Discovery from Parallel and Distributed Databases*, 2000.
10. S.I. McClean, B.W. Scotney, and S. Robinson. Conceptual clustering of heterogeneous gene expression sequences. *Artificial Intelligence Review*, 20:53–73, 2003.
11. G.S. Michaels, D.B. Carr, M. Askenazi, S. Fuhrman, and X. Wen et al. Cluster analysis and data visualisation of large-scale gene expression data. *Pacific Symposium on Biocomputing*, 3:42–53, 1998.
12. T.M. Mitchell. *Genetic Algorithms in Search, Optimization, and Machine Learning*. McGraw-Hill, 1997.
13. M. Molla, P. Anreae, J. Glasner, F. Blattner, and J. Shavlik. Interpreting microarray expression data using text annotating the genes. *Information Science*, 146:75–88, 2002.
14. P. Smyth. Clustering sequences with hidden markov models. In Jordan MI Petsche T Mozer MC, editor, *Advances in Neural Information Processing*. MIT Press, 1997.

Analysis of Large-Scale mRNA Expression Data Sets by Genetic Algorithms

Chia Huey Ooi[1] and Patrick Tan[1,2]

[1] National Cancer Centre, Singapore 169610, Republic of Singapore.
E-mail: `Chia.Huey.Ooi@infotech.monash.edu.au`

[2] National Cancer Centre, Defence Medical and Environmental Research Institute, Republic of Singapore.
E-mail: `cmrtan@nccs.com.sg`

Summary. DNA microarray experiments typically produce large-scale data sets comprising thousands of mRNA expression values measured across multiple biological samples. A common problem in the analysis of this data is the 'curse of dimensionality', where the number of available samples is often insufficient for reliable analysis due to the large number of individual measurements made per sample. *Genetic algorithms* (GAs) are a promising approach towards addressing this problem. Here, several examples from the scientific literature are discussed to illustrate the various ways GAs have been applied to the analysis of mRNA expression data sets, specifically in the area of cancer classification.

1 Introduction

One of the most exciting developments in recent *functional gemomics* research has been the application of gene expression profiling technology to the study of oncology and the specific problem of cancer classification. Being a complex disease, cancer happens when a critical number of changes have been accumulated in the same cell over a period of time. Furthermore, matters become more complicated when we consider that there are many different types (not to mention different subtypes within a single type!) of cancer, each having arisen from different combinations of changes.

At present, the clinical and molecular heterogeneity inherent to many cancers presents a significant challenge towards the successful diagnosis and treatment of the disease—it is well-known that individual cancers can exhibit tremendous variations in clinical presentation, disease aggressiveness, and treatment response, suggesting that these clinical entities may actually represent a conglomerate of many different and distinct cancer subtypes. In contrast to conventional techniques currently being used by clinical histopathologists such as light microscopy and *immunohistochemistry* (IHC), the use of expression profiles to classify tumors may convey several advantages. First, it has been shown that expression profiles can define clinically relevant subtypes of cancer that have previously eluded more conventional approaches

W. Dubitzky and F. Azuaje (eds.), Artificial Intelligence Methods and Tools for Systems Biology, 51–66.

[1, 6]. Second, in contrast to the use of single molecular markers, the ability to monitor the expression levels of multiple genes in a simultaneous fashion can often provide a useful insight into the activity state of clinically significant cellular and tumorigenic pathways. Third, depending on the scoring pathologist, results from conventional IHC may sometimes be misleading due to the presence of isolated aberrant regions on the tissue section. In contrast, because expression profiles are usually derived from the bulk of the tumor, they may better represent the overall collective biology of the composite tumor.

Reflecting this promise and the active nature of the field, a substantial body of work now exists describing various analytical strategies exploiting mRNA expression profiles for cancer classification. Typically, the expression data set is analyzed using a wrapper gene selection-classifier training approach to identify genes whose expression is selectively enriched in one tumor type ('A') and not another ('B') (Fig. 1). This is usually achieved as follows: for an expression data set comprising of M samples (arrays) and N genes (features), the samples are first divided into a training set (M_T samples) and test set (M_I samples). Feature selection and classifier training is performed using the training set samples, and the accuracy of the trained classifier is determined by testing its ability to classify the test set samples. Importantly, samples belonging to the test set should be completely 'hidden' from the initial gene selection process performed using the training set. In addition to the gold standard of an independent test, *V-fold cross-validation* or bootstrap methods can also be employed on the training set to obtain an unbiased estimate of sampling error [3]. With the former, the training set is first randomly split into V subsets. A subset is singled out, and the feature selection process is repeated on the remaining $V - 1$ subsets before attempting to classify the singled-out subset. This is repeated for all subsets, and the total error rate is recorded as the estimate of error, E_C (Note that this becomes a *leave-one-out cross-validation* (*LOOCV*) assay when $V = MT$). In bootstrapping, instead of having V subsets that are exclusive of each other, B subsets of size M_T are defined, where each subset consists of samples that are resampled from the original training set with replacement. Each subset is trained to produce a predictor set, which is then used to classify all the training samples. E_C is then computed based on the error rate from each subset. Interested readers are referred to [10] for further details on error estimation.

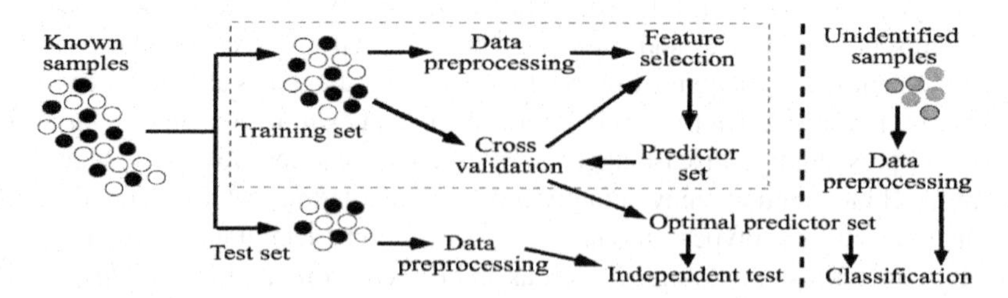

Fig. 1. Wrapper-approach gene selection scheme (left) for tumor classification (right).

1.1 The challenge of high dimensionality for cancer classification

Although it is technically possible to build an apparently accurate classifier for cancer using the expression data from every single gene captured on a microarray, for widespread clinical use it may not be feasible to use a diagnostic test that requires monitoring the expression levels of thousands of genes [31]. Thus, in addition to prediction accuracy, another important consideration for the ideal cancer classifier is that it should consist of as few genes as possible while preserving good classification accuracy. This may facilitate the development of simpler, faster and cheaper diagnostic microarrays containing probes for only 'key' classifier features. There thus lies the need to select, among the thousands of measured genes on a microarray, the handful that will form a good predictor set. In this context, it is worth noting that the heuristic 'goodness' of a predictor set can often be measured by the extent to which it obeys the hypothesis that good predictor sets should contain features highly correlated with the specific class distinction, and yet uncorrelated with one other [15].

A major challenge in this regard, however, is the 'curse of dimensionality' that arises as a consequence of the high-dimensionality nature of microarray data [20]. Typically, the design of a classifier should utilize at least 10 times as many training samples per class as the number of features [19]. However, this criteria is extremely difficult, if not impossible, to satisfy for most microarray experiments, where the typical mRNA expression data set can contain at least 2 000 features and at most 200 samples, which, even in case of 2-class data sets, barely fulfills the aforementioned criteria. Predictably, the situation is worsened when one considers multi-class data sets, where the number of classes can extend to five or more.

The high dimensionality of mRNA expression data sets also presents a specific challenge when one considers the requirement that a predictor set for tumor classification should ideally consist of a small number of features. A suitable analogy would be 'looking for a needle in a haystack', as one would need to select, from all available measured features, the appropriate combination of 10 to 100 to be presented to the classifier that maintains a reasonably good classification accuracy. As an illustration, the solution space for a 20-gene predictor set drawn from a data set of 2 000 genes would consist of 3.9×10^{47} possible candidates, and this space would exponentially increase if a range of predictor set sizes was tested. Because of these difficulties, there is thus a need to develop algorithms that can analyze these large solution spaces in a rapid and parallel manner. Genetic algorithms, which are now described, may present a possible analytical approach that can be effectively applied to this problem.

2 Genetic algorithms—a brief introduction

Genetic algorithms (GAs), first introduced by John Holland in the mid-1970s, are randomized search and optimization techniques that derive their working principles from the processes of evolution and natural genetics. In non-biological applications such as engineering, network designs and traditional machine learning, GAs have already been well-established as heuristic optimization tools. In particular, because they are aided by large amounts of implicit parallelism [14], GAs are capable of searching for optimal or near-optimal solutions on complex and large spaces of possible solutions. This analytical capability thus makes GAs highly suited to the complexity of a typical mRNA expression data set. Some specific advantages that GAs bring to the analysis of mRNA expression data sets include:

1. GAs have been proven as an effective feature selector [26],
2. GAs are easily adaptable as they can be used with any single or combined classifiers,
3. GAs are flexible in algorithm design, as key parameters can be adjusted by the user to achieve optimal results.

There are 5 distinct stages in a typical GA: *initialization*, *evaluation*, *selection*, *crossover* and *mutation*. These do not occur linearly, but rather in a generation loop as shown in Fig. 2.

In the GA, the evolutionary process is simulated by first initializing a random population of 'individuals'. Each 'individual' represents a potential solution to the problem at hand, and is usually constructed as a *string* of *variables* or '*genes*' [17], which represent the specific attributes or parameters of the solution.

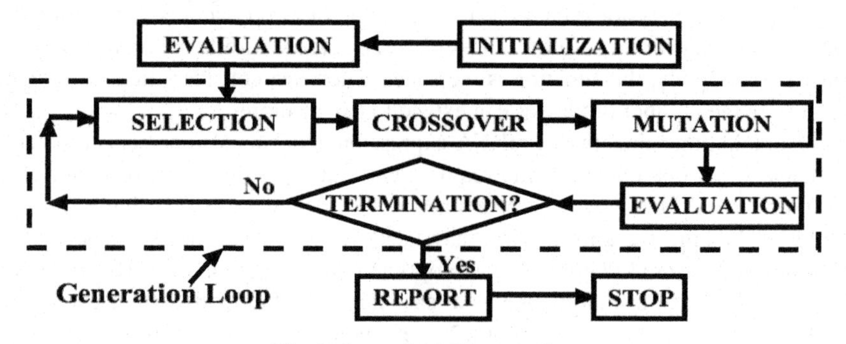

Fig. 2. A standard GA procedure.

To evaluate the specific 'goodness' of a solution represented by an individual, a fitness function is applied to each member of the population in order to calculate the fitness value of each individual. In the selection stage, a 'mating pool' is created whose members are derived from the current generation. Echoing the concept of 'survival of the fittest', individuals with higher fitness values are assigned higher probabilities of 'mating' by having more copies of themselves in the mating pool. In the mating pool, crossover and mutation operations are performed to create a new and diverse set of solutions. In crossover, two individuals are randomly chosen from the mating pool and a crossover operation is applied on the selected string pair with a fixed probability p_c (the expected number of crossovers occurring per individual per generation), generating two offspring strings through the exchange of genetic information between the two parents. This probabilistic process is repeated until all parent strings in the mating pool have been considered. Mutation operations are also applied at a pre-determined rate, upon offspring strings produced from crossover. The most common application is *point mutation* where the value of a variable at a probabilistically determined position along the offspring string is changed at probability p_m (the expected number of mutations to occur per 'gene' in a generation).

The resulting set of derived individuals now serves as a new population upon which a new generation loop of evaluation, selection, and crossover/mutation is performed. These generation cycles are reiteratively executed until a pre-specified termination criterion is fulfilled. Typical termination criteria can either be:

1. The appearance of an individual or a set of individuals which satisfy the required fitness value,
2. A situation where no significant improvement of average or best fitness value of the population is achieved after the duration of a pre-determined number of consecutive generations (i.e., convergence has been attained),
3. The algorithm has been reiterated for a pre-determined maximum number of generations.

In essence, the GA allows relatively fitter individuals to thrive compared to their less fit counterparts, as the former are able to perpetuate their genes across subsequent generations at the expense of the latter. Crossover and mutation mechanisms also make it possible for more 'optimized' individuals to arise from fit individuals of the previous generation. Finally, selection mechanisms ensure that the newer, fitter individuals are preserved in the following generation, resulting in an increase of the average fitness of the total population.

2.1 Implementing a GA for feature selection

This section presents a series of simple GA designs for feature selection that can be used for cancer classification. String design is perhaps the most important component of a GA. One intuitive string design that could be used in feature selection would be a *binary-encoding*-based design, where each string contains N variables and each variable represents a gene drawn from a microarray data set of N genes. For each variable, a value of 0 or 1 can be assigned, where a value of 1 indicates that gene i is to be included in the predictor set to be presented to the classifier, and a value of 0 means the exclusion of gene i (Fig. 3a).

One disadvantage of this basic string design is that it is a rather inefficient coding technique, since each string will contain as many variables as the total number of genes in the microarray data set. Other more efficient string designs can also be used to describe a predictor set. For example, in a whole number encoding string, an index number is assigned to each gene in the data set $(1, 2, ..., N)$ and the string is composed of variables ranging from 0 to N (Fig. 3b), where a variable with the value of 0 represents no gene. Therefore instead of representing the total number of genes in the data set, N, the actual string length, L represents a user-specified upper limit to the number of genes in the predictor set, R_{max}. Genes whose index numbers

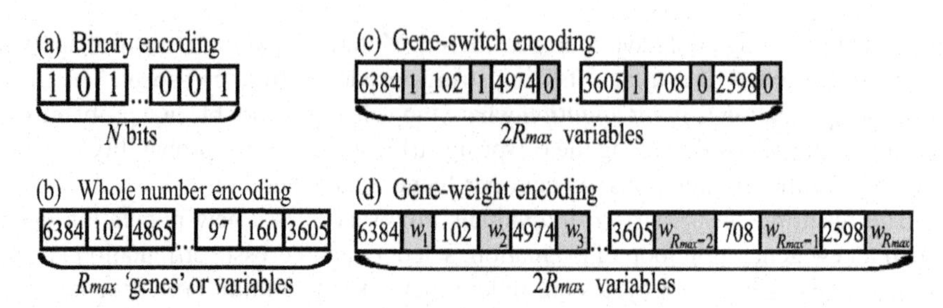

Fig. 3. Some feasible string designs for feature selection in cancer classification ranging from the simplistic binary encoding to the more complex gene-weight encoding.

are contained within a string are included in the predictor set. Due to the possible presence of variable(s) with the value of 0, in this scheme a predictor set can contain anywhere from 0 to R_{max} genes.

Now, it is known that the number of possible string permutations is d^L (when there are d possible values for each of the variable in the L-variable string). Therefore the solution space that needs to be searched in case of the binary encoding is in the order of 2^N, while for the whole number encoding it is in the order of $N^{R_{max}}$. Since according to the definition of a good predictor set (discussed in Section 1.1), $R_{max} \ll N$, the solution search space is smaller than the one for the binary encoding.

A combination of both the binary and whole number designs produces the *gene-switch encoding* (Fig. 3c), where a gene index and its 'switch' ('0' for 'off', '1' for 'on') are placed side-by-side along the string. An extension of this idea is the gene-weight encoding string, where the 'switch' for each gene index is replaced as a weight, w_{G_i}, which represents the contribution of gene i to the predictor set (Fig. 3d).

Once a string design has been determined, a starter population of P individuals can then be randomly initialized. P typically ranges from 50 to the order of 10 000, depending upon computational limitations and the required diversity of the population.

A second key component of the GA is the structure of the objective or fitness function, which measures the 'goodness' of an individual S_i based on pre-specified criteria. In the specific case of cancer classification, one simple measure of 'goodness' might be the ability of a predictor set encoded by a particular string to deliver good classification accuracies. The situation may be more complicated, however, when multiple criteria are considered in determining the overall 'goodness' of an individual. For multi-objective GAs, the overall fitness can be expressed in the vector form $\mathbf{f}(S_i) = \langle f_1(S_i), \ldots, f_K(S_i) \rangle$ where the scalar function f_k represents the k-th objective. For example, besides attaining good classification accuracies, one might also want the GA to select the individual containing the least number of genes. In this case the objectives could at least be: (1) maximize training classification accuracy and (2) minimize the number of selected genes. This could be represented by:

$$f_1(S_i) = 100 - E_C \tag{1}$$

$$f_2(S_i) = 1/R \tag{2}$$

where E_C is the cross-validation error rate (and hence in this case $f_1(S_i)$ would be the cross-validation accuracy). Thus for an individual S_i the fitness value increases with accuracy, but faces penalization if the predictor set size, R is too large.

If a plain aggregation method is used, the fitness function can then be expressed in a scalar form

$$f(S_i) = \sum_{k=1}^{K} = w_k f_k(S_i) \tag{3}$$

where w_k represents the weight set to each corresponding objective $f_k(S_i)$.

In selecting the individuals that will enter the mating pool, two steps are involved: fitness assignment, followed by actual selection. Fitness assignment can be performed using different operations. In proportional fitness assignment [12], the transformed fitness $F(S_i)$ is obtained by dividing the raw fitness $f(S_i)$ by the sum of $f(S_i)$ of the population ($\sum f(S_j)$ for $j = 1, 2, \ldots, P$). However, as one may encounter various scaling problems with this method, an alternative methodology is the more robust rank-based fitness assignment [4], where the population is first sorted based on their raw fitness values, $f(S_i)$ and each individual is then assigned an adjusted fitness value $F(S_i)$ purely based on its ranked position.

The next step of actual selection can be implemented using a number of methods. In *stochastic universal sampling* (SUS, Fig. 4a), the population is represented as a single contiguous line, and each individual in the population is represented as a segment on that line. The length of each segment is based upon the fitness of the individual, i.e., an individual with higher fitness would be assigned a longer segment. N equally spaced pointers are placed over the line, where the position of the first pointer is chosen randomly in the range $[0, 1/P]$. The number of copies of a particular individual S_i entering the mating pool is equal to the number of pointers that fall within S_i's segment. This selection method gives zero *bias* and minimum *spread* [5], which are desirable in a selection method. Bias is defined as the absolute difference between an individual's transformed fitness $F(S_i)$ value (as defined in the proportional fitness assignment method) and its expected probability of reproduction, while spread is the range of possible values for the number of offspring assigned to an individual [5].

The *roulette wheel selection* (RWS, or *stochastic universal sampling with replacement*, Fig. 4b) also uses a single contiguous line to represent the population. In this case, a single pointer is placed at a random position along the line P times. In each placement, a copy of the individual S_i on whose segment the pointer falls enters the mating pool. Although this is a simpler selection scheme, it does not guarantee minimum spread [5].

These aforementioned methods are best used in the context of a single scalar fitness function. If the fitness function is in vector form, other selection methods can

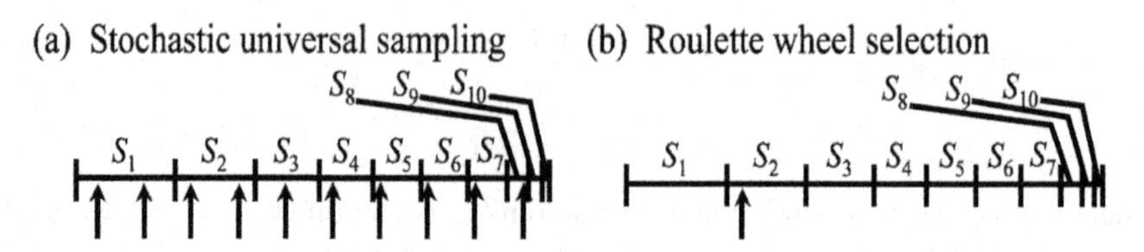

Fig. 4. Two of the most conventional stochastic selection methods used in GA.

be used. We briefly discuss two selection methods that have been established for multi-objective GAs, based upon non-Pareto and Pareto-ranking approaches. The *vector evaluated genetic algorithm* (VEGA) [28] performs selection in a non-Pareto way. The population is divided into K groups of equal size, where K represents the number of criteria/objectives. In the k-th group proportional fitness selection is implemented based on the scalar function $f_k(S_i)$. The fittest individuals from the K groups are then mixed to create a combined mating pool where crossover and mutation operations are applied. While VEGA provides a robust selection technique for high order optimization problems when little *a priori* knowledge is available to guide the search, it faces the problem of *speciation* [28], where the population splits into groups of species particularly strong in each of the components of $\mathbf{f}(S_i)$. This tendency is due to the fact that VEGA implicitly performs linear combination of the objectives [11].

In *Pareto-based approaches*, each individual in the population is compared to every other in order to identify non-dominated individuals. An individual is considered to be 'non-dominated' when there does **not** exist any other individual in the population that is (1) equal or superior to it in **all** criteria; and (2) superior to it in at least one criterion. The 'non-dominated' individuals are then isolated from the population and moved into the first non-dominated *front*. (A front is a set of individuals with the same rank of non-domination.) The remaining individuals in the population are again evaluated, and the now non-dominated individuals are similarly moved into the second non-dominated front. This process of evaluation and isolation is repeated until all individuals in the population have been placed in Pareto-based fronts. All members of one front receive the same rank, and members of the first front have the highest rank [12]. After ranking, fitness values are then assigned linearly from the best to the worst individual. The *non-dominated sorting genetic algorithm* (NSGA) [29], discussed in the next section, uses this ranking approach.

After selecting the individuals comprising the mating pool, crossover and mutation operations can now be performed. Two common types of crossover are shown in Fig. 5. The reader is referred to [16] for further reading. In *one-point crossover*, a common cut point is randomly chosen along the strings, producing segments that are then exchanged to produce two genetically different offspring. In *uniform crossover*, the elements of a randomly generated template determine the genes a child receives from each parent.

String design is an important factor in deciding the appropriate type of crossover function to use. If the genes are arbitrarily arranged along the string (Fig. 3a and

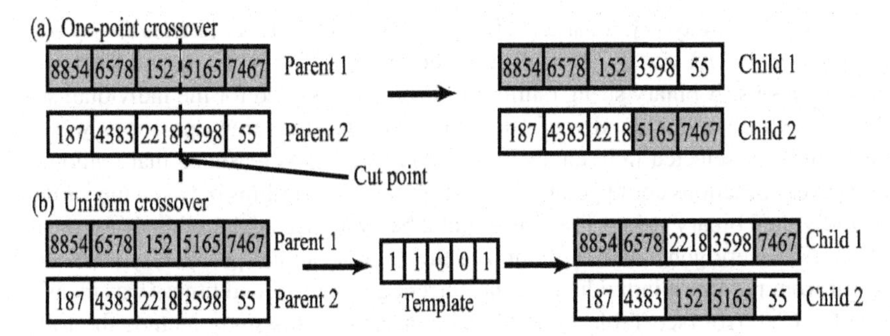

Fig. 5. Two disparate variants of crossover operators used in GA. At one extreme is the one-point crossover, while the uniform crossover (which is essentially a multi-point crossover) stands at the other end.

3b), uniform crossovers should perform comparably [30], if not better than one-point crossover, since *disruption* [17, 21] is not a factor in this case. However, if the variables representing the genes are ordered along the string based on external information, such as their physical genomic location, one-point or custom-designed crossovers are preferred over uniform crossover. The choice of mutation operator can also be affected by the string design. For binary encoding strings, the most common mutation operator is the bit-flip operator, where the bit of the 'gene' selected for mutation is simply flipped ('1' becomes 0, and vice versa). In the case of whole number encoding strings, random mutation can be used, where the value of a variable (i.e., the gene index number) that has been chosen for mutation is replaced by a randomly chosen value between 0 and N.

Finally, the reconstitution of a new population for the next generation can employ various types of replacement schemes. In *complete replacement* schemes, all the resulting offspring from the genetic operations and none of the original parents populate the next generation. In *elitist replacement* schemes, both the parents and offspring are ranked by their fitness values, and the top P ranking individuals enter the next generation. Thus, fit parents are retained. The advantage of such elitist schemes over complete replacement schemes is that good individuals are not lost forever if they produce inferior offspring. A disadvantage of the elitist approach, however, is the increased risk of leading the GA into a local rather than global optimum.

2.2 Some examples from the literature

There are a number of reported studies (see Table 1) in the literature describing GA-based strategies for feature selection and cancer classification. The performance of these approaches was evaluated by their ability to correctly classify a number of publicly available microarray data sets, including those for colon cancer [2], leukemia [13], lymphoma [1], human tumor cell lines of diverse tissue origins (NCI60) [27] and multiple primary tumor types ('Global Cancer Map', or GCM) [25]. The first three data sets represent 2-class classification problems, while the last two (NCI60 and GCM) are multi-class data sets (9 classes for NCI60 and 14 classes for GCM).

In the first study [7], a variant of GAs referred to as EDA (*estimation of distribution algorithms*) was used to select predictor genes for the colon and leukemia 2-class data sets. A binary string (similar to Fig. 3a) was used for the individual string design with a population size of 100. Crossover and mutation operations were not performed on selected individuals in this study, as it was reasoned that randomized crossover operations could potentially disrupt the relationships between the variables on individual binary strings [18] (see point 2 below). Instead, individuals for the next generation were generated by sampling the L-dimensional probability distribution learnt from the fittest individuals of the current generation, and fitness was measured by LOOCV error (see Table 1). Potential concerns in this study include the following:

Table 1. GA-based studies in tumor classification.

No.	Study	Data Sets	Cross-Validation	Test	No. of Genes	Classifier
1	EDA [7]	Colon	98.4	NA	6	naive
		Leukemia	100	NA	8	Bayes (NB)
2	GA/KNN [22]	Colon	95–100	94.1	50	k-nearest
		Lymphoma	91–100	84.6	50	neighbor (k-NN)
		Colon	90	NA	14	
3	MOEA [23]	Leukemia	97	NA	16	weighted voting
		Lymphoma	94	NA	18	
4	GA/MLHD [24]	NCI60	85.4	95	13	linear
		NCI60	79.3	86	32	discriminant
		Leukemia	100	100	3	weighted
		Lymphoma	100	100	5	voting: 2-class
5	NSGA-II [9]	Colon	100	100	6	problems,
		NCI60	92.7	90	37	one-vs-all
		GCM	86	80	12	multi-class

1. The authors did not establish a test set that was separate from the training process. Instead, for both the leukemia and the colon data sets, all the samples were used in the training set. Hence, there was no way of independently estimating the accuracy of the best predictor sets obtained by the EDA,
2. The selection stage of the EDA assumes that the variables represented in a particular string are independent of each other. This, however, may contradict the prior concern (described above) regarding crossovers disrupting relationships between individuals, which suggests the existence of interdependencies between the variables.

In contrast to [7], the GA/KNN (k-nearest neighbors) study [22] used a string design consisting of actual gene indices rather than a binary string to represent a potential predictor set. The size of a predictor set was not encoded within the string, and various string length values, from $L = 5$ up to 50 were independently tested in separate runs. Using a population size of 100, 6 discrete predictor set sizes (of 5, 10, 20, 30, 40 and 50 genes per set) were separately considered. Similar to the EDA study, the fitness function was the cross-validation accuracy. Mutation operations were performed as follows: One to five 'genes' per individual were randomly chosen and replaced with other 'genes' that were not already present in the predictor set. Crossover operations were not performed in this study. Instead of specifying a fixed number of generations, a termination criterion of 31/34 and 38/40 cross-validation accuracy was utilized for the lymphoma and colon data sets respectively.

One interesting aspect of this study, however, was that unlike typical GAs designs where an optimal predictor set is typically derived from one or a few runs, this study adopted a 'pooling' approach to identify a consensus predictor set that could be ultimately used by a k-NN classifier to classify the test set. Specifically, all the individual genes were first ranked by their frequency of selection based upon the top 10 000 predictor sets generated by running the GA on the training set using different values of L. For each value of L, the N top-ranked genes were then picked as predictor genes, and were used to classify the test sets through the k-NN method. This study found that in general for all values of L: (1) the windows for the best test set accuracy were achieved using classifier sizes of < 250 and (2) both test and training set classification accuracies decreased with increasing N beyond 250, demonstrating the essentiality of good predictor set selection in the classification process.

The challenge of identifying a predictor set that fulfills more than one requirement was recently addressed in a MOEA (*multi-objective evolutionary algorithm*) study [23], where the investigators made full use of the power of multi-objective GAs by attempting to simultaneously optimize 3 parameters: the number of misclassified training samples (F_1), the predictor set size (F_3) and the class bias (F_2), which reflects the potential bias of a predictor set due to unequal class sizes. A binary string design (similar to Fig. 3a) was used, and the predictor set size was controlled during population initialization by randomly generating strings where the total number of variables assigned a '1' was constrained between two thresholds τ_1 to τ_2, τ_1 and τ_2 being user-specified constants.

To evaluate the predictor sets, two subsets (size ratio 7:3) were randomly formed from the data set. The larger set, together with the predictor set of the individual, was used to train a weighted voting classifier which was then challenged by the members of the smaller set, which was hidden from the classifier. The fitness values were then calculated based on the individual's non-domination level (number of individuals not dominated by it).

A final predictor set was obtained after 10 runs of the MOEA classifier, with a crossover rate of 0.6 per individual per generation, mutation rate 0.001, population size 500 and 200 000 generations per run. Since the aim was to obtain a single predictor set, not the whole best Pareto front, simple aggregation of the F_1, F_2 and F_3 objectives was used to determine the optimal individual from members of the final or best Pareto front. In simple aggregation, the objectives are combined into a higher scalar function that is used as the criterion of decision making. The optimal predictor sets were tested by performing LOOCV on the 3 data sets used in this study in order

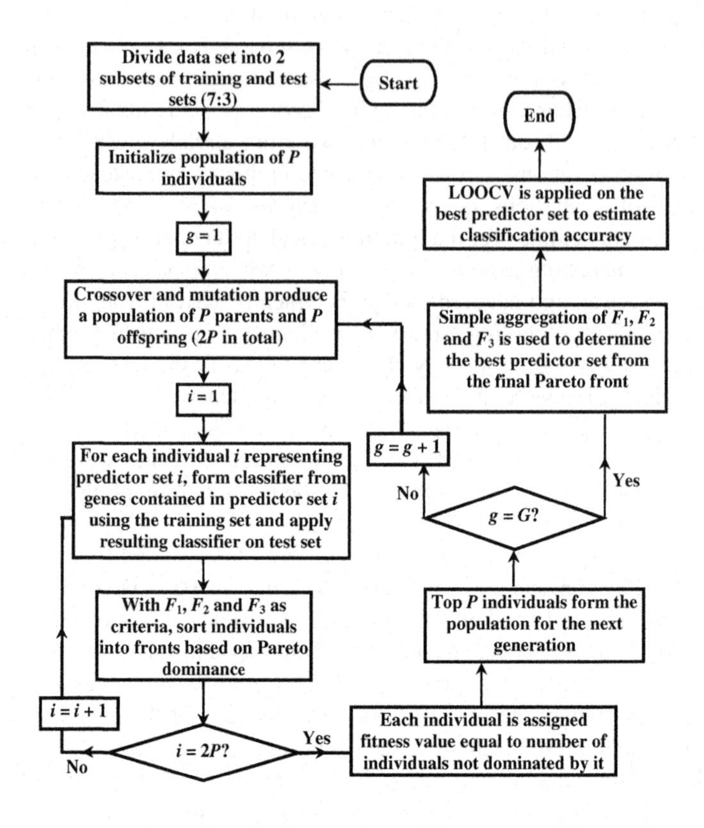

Fig. 6. A Summary of the MOEA approach. G denotes the maximum number of generations in an MOEA run, while P is the pre-specified population size.

to obtain the results shown in Table 1. A summary of this approach is presented in Fig. 6.

Besides 2-class data sets, at least two studies have considered how GAs could be applied to the problem of classification in the context of multiple classes. Multi-class data sets were first considered in the GA/MLHD study [24], which combined a simple GA design with a maximum likelihood discriminant (MLHD) classifier. The fitness function was a single scalar function defining classification accuracy as the only parameter to be maximized. Two multi-class data sets were considered—the 9-class NCI60 and the 14-class GCM. In analyzing the NCI60 data set, the predictor set size, R was coded as the initial character of the individual string design and hence not included in the fitness function. The remainder of the strings contained integers $[1, N]$, representing the index of a gene. To control the predictor set size, a method similar to [23] was adopted where the population was initialized such that R was between pre-determined constants R_{min} and R_{max}, and string length was fixed at $R_{max} + 1$. For the GCM data set, a slightly different string design was employed (similar to Fig. 3c), where the predictor set size could be anywhere from 0 to a pre-determined maximum, R_{max}. Interestingly, it was found that the value of R_{max} did not appear to play a significant role in influencing the ultimate size of the final predictor sets. For example, the size of optimal predictor sets converged around 30 genes by generation 100 or so, regardless of whether R_{max} was set to 60 or 100. Analysis of the NCI60 data set used a population size of 100 and set the maximum number of generations to 100, while for the GCM data set the population size was 30 and the maximum number of generations, 120. Various combinations of crossover (uniform, one-point) and selection operators (RWS, SUS) were tested in different runs. The optimal predictor set size was discovered to be between the range $[11, 15]$ for the 9-class NCI60 and $[1, 60]$ for the 14-class GCM) after various runs utilizing different size ranges (from $[1, 5]$ to $[26, 30]$), varying crossover (0.7 to 1.0) and mutation rates (0.0005 to 0.02). It is worth noting that although the linear discriminant classifier used in this method tends to select uncorrelated genes into optimal predictor sets, thereby complying with the hypothesis defining a 'good' predictor set [15], the necessity to invert the covariance matrix in the discriminant analysis comes at a fairly high computational cost. Developing new methods to correct this disadvantage will be an important issue for future research.

Finally, a separate study [9] also addressed cancer classification problems for both multi- and 2-class data sets using the NSGA-II (improved version of the NSGA). Advantages of the NSGA-II approach include its modest computational complexity and an efficient diversity preservation method [8]. A binary string (Fig. 3a) design was adopted. To reduce the length of each string and lower the computational cost, some of the more complex data sets were preprocessed by removing certain genes based upon the differences between their maximum and minimum expression values.

For each of the 2-class data sets analyzed, the NSGA-II was run with a population size 1 000 for 1 000 generations, with one-point crossover at a probability of 0.8 and a mutation rate of 0.0005 (0.001 for the colon data set). With the multi-class data sets, the mutation probability was decreased to 0.0001, while the maximum number of generations was increased to 2 000 in case of the GCM data set (which comprises a larger number of samples, genes and classes to consider). Since the NSGA-II is a multi-objective GA, 3 objectives were defined: f_1: predictor set minimization, f_2: training error minimization, and f_3: test error minimization. Notably, this is the first study in which a GA-based tumor classifier was applied to all five popular expression data sets.

3 Conclusions—and the future

The few examples cited here demonstrate that the highly parallelized searching capabilities of GAs can be effectively used as powerful optimization tools to create optimal predictor sets for cancer classification, both for 2-class and multi-class mRNA expression data sets. Nevertheless, much work remains to be done before the true applicability of GAs to cancer classification can be fully appreciated. One of the most important issues is the problem of *selection bias*, which is now recognized as a common flaw in many microarray-based cancer classification studies (involving both GA and non-GA based methodologies). In many of these studies where LOOCV was used as a surrogate index of classification accuracy, in the majority of these cases the initial feature selection process utilized all samples (rather than all samples minus one), which can also lead to significant overfitting [3]. In addition, for a test set to provide an unbiased measure of classification performance by a predictor set, that test set must also be completely segregated from even the initial feature selection process. In the specific case of GAs, another potential concern is that for all the methodologies described in this chapter employing a test set (with the exception of the GA/KNN study), the classification accuracy of the test set was also incorporated into the fitness function—the effects of this with regard to potential overfitting remains an important issue to be addressed.

Another relatively unexplored area lies in the potential use of individuals where the string length is variable and not fixed. All the studies discussed in this chapter employed the fixed-length 'chromosome' concept, which greatly simplifies the implementation of the crossover operator. However, it will be interesting to study the possible contribution of variable-length string designs to GA-based methods for feature selection. In such a scenario, individuals with varying lengths (i.e. representing predictor sets of various sizes) could compete with one another as well. In closing, we believe that the full extent of the potential contributions of GAs to feature selection remains to be fully exploited, particularly in the specific area of expression genomics and cancer classification.

Acknowledgments

PT thanks Hui Kam Man for his support and encouragement. This work was funded by a core grant to PT from NCC

References

1. A.A. Alizadeh, M.B. Eisen, R.E. Davis, C. Ma, and I.S. Lossos et al. Distinct types of diffuse large b-cell lymphoma identified by gene expression profiling. *Nature*, 403:503–511, 2000.
2. U. Alon, N. Barkai, D.A. Notterman, K. Gish, and S. Ybarra et al. Broad patterns of gene expression revealed by clustering analysis of tumor and normal colon tissues probed by oligonucleotide arrays. *PNAS*, 96:6745–6750, 1999.
3. C. Ambroise and G.J. McLachlan. Selection bias in gene extraction on the basis of microarray gene-expression data. *PNAS*, 99:6562–6566, 2002.
4. J.E. Baker. Adaptive selection methods for genetic algorithms. In J.J. Grefenstette, editor, *Proc. Int. Conference on Genetic Algorithms and their Application*, Hillsdale, NJ, 1985. Lawrence Erlbaum Assoc.
5. J.E. Baker. Reducing bias and inefficiency in the selection algorithm. In J.J. Grefenstette, editor, *Proc. 2nd Int. Conference on Genetic Algorithms and their Application*, Hillsdale, NJ, 1987. Lawrence Erlbaum Assoc.

6. M. Bittner, P. Meltzer, Y. Chen, Y. Jiang, and E. Seftor et al. Molecular classification of cutaneous malignant melanoma by gene expression profiling. *Nature*, 406:536–540, 2000.
7. R. Blanco, P. Larraaga, I. Inza, and B. Sierra. Selection of highly accurate genes for cancer classification by estimation of distribution algorithms. In S. Quaglini, P. Barahona, and S. Andreassen, editors, *Proc. 8th Conference on Artificial Intelligence in Medicine in Europe*, Heidelberg, 2001. Springer.
8. K. Deb, A. Pratap, S. Agarwal, and T. Meyaxivan. A fast and elitist multiobjective genetic algorithm: Nsga-ii. *IEEE Transactions on Evolutionary Computation*, 6, 2002.
9. K. Deb and A.R. Reddy. Classification of two and multi-class cancer data reliably using multi-objective evolutionary algorithms. Technical report, Indian Institute of Technology, Kanpur Genetic Algorithms Laboratory, Indian Institute of Technology, Kanpur, 2003. Retrieved November 2003 from: http://www.iitk.ac.in/kangal/papers/k2003006.pdf.
10. B. Efron and R. Tibshirani. Cross-validation and the bootstrap: Estimating the error rate of a prediction rule. *J Am Stat Assoc*, 92:548–560, 1997.
11. C.M. Fonseca and P.J. Fleming. An overview of evolutionary algorithms in multiobjective optimization. *Evolutionary Computation*, 3, 1995.
12. D.E. Goldberg. *Genetic Algorithms in Search, Optimization, and Machine Learning*. Addison-Wesley, Reading, Massachusetts, 1989.
13. T.R. Golub, D.K. Slonim, P. Tamayo, C. Huard, and M. Gaasenbeek et al. Molecular classification of cancer: Class discovery and class prediction by gene expression monitoring. *Science*, 286:531–537, 1999.
14. J.J. Grefenstette and J.E.Baker. How genetic algorithms work: A critical look at implicit parallelism. In J.D. Schaffer, editor, *Proc. 3rd Int. Conference on Genetic Algorithms*, San Mateo, CA, 1989. Morgan Kaufmann.
15. M.A. Hall and L.A. Smith. Practical predictor set selection for machine learning. In C. McDonald, editor, *Proc. Australasian Computer Science Conference*, Singapore, 1998. Springer.
16. R.L. Haupt and S.E. Haupt. *Practical Genetic Algorithms*. John Wiley and Sons, New York, 1998.
17. J. Holland. *Adaptation in Natural and Artificial Systems*. MIT Press, Cambridge, MA, 2nd edition, 1992.
18. I. Inza, P. Larraaga, and B. Sierra. Predictor set selection by bayesian networks: a comparison with genetic and sequential algorithms. *International Journal of Approximate Reasoning*, 27, 2001.
19. A.K. Jain and B. Chandrasekaran. Dimensionality and sample size considerations in pattern recognition practice. In P.R. Krishnaiah and L.N. Kanal, editors, *Handbook of Statistics*, volume 2, pages 835–855. North-Holland, Amsterdam, 1982.
20. A.K. Jain, R.P.W. Duin, and J. Mao. Statistical pattern recognition: A review. *IEEE Transactions on Pattern Analysis and Machine Intelligence*, 22:4–37, 2000.
21. K.A. De Jong. *An Analysis of the Behaviour of a Class of Genetic Adaptive Systems*. PhD thesis, University of Michigan.
22. L. Li, C.R. Weinberg, T.A. Darden, and L.G. Pedersen. Gene selection for sample classification based on gene expression data: study of sensitivity to choice of parameters of the ga/knn method. *Bioinformatics*, 17:1131–1142, 2001.
23. J. Liu and H. Iba. Selecting informative genes using a multiobjective evolutionary algorithm. In D. Fogel, editor, *Proc. Congress on Evolutionary Computation*, Piscataway, NJ, 2002. IEEE Service Center.
24. C.H. Ooi and P. Tan. Genetic algorithms applied to multi-class prediction for the analysis of gene expression data. *Bioinformatics*, 19:37–44, 2003.

25. S. Ramaswamy, P. Tamayo, R. Rifkin, S. Mukherjee, and C.H. Yeang et al. Multi-class cancer diagnosis using tumor gene expression signatures. *PNAS*, 98:15149–15154, 2001.
26. M.L. Raymer, W.F. Punch, E.D. Goodman, L.A. Kuhn, and A.K. Jain. Dimensionality reduction using genetic algorithms. *IEEE Transactions on Evolutionary Computation*, 4, 2000.
27. D.T. Ross, U. Scherf, M.B. Eisen, C.M. Perou, and C. Rees et al. Systematic variation in gene expression patterns in human cancer cell lines. *Nature Genetics*, 24:227–235, 2000.
28. J.D. Schaffer. Multiple objective optimization with vector evaluated genetic algorithms. In J.J. Grefensttete, editor, *Genetic Algorithms and Their Applications: Proc. 1st Int. Conference on Genetic Algorithms*, Hillsdale, NJ, 1985. Lawrence Erlbaum Assoc.
29. N. Srinivas and K. Deb. Multiobjective optimization using non-dominated sorting in genetic algorithms. *Evolutionary Computation*, 2, 1995.
30. G. Syswerda. Uniform crossover in genetic algorithms. In J.D. Schaffer, editor, *Proc. 3rd Int. Conference on Genetic Algorithms*, San Mateo, CA, 1989. Morgan Kaufmann.
31. H.P. Zhang, C.Y. Yu, B. Singer, and M.M. Xiong. Recursive partitioning for tumor classification with gene expression microarray data. *PNAS*, 98:6730–6735, 2001.

A Data-Driven, Flexible Machine Learning Strategy for the Classification of Biomedical Data

Rajmund L. Somorjai[1], Murray E. Alexander[1], Richard Baumgartner[1], Stephanie Booth[2], Christopher Bowman[1], and Aleksander Demko[1], Brion Dolenko[1], Marina Mandelzweig[1], Aleksander E. Nikulin[1], Nicolino J. Pizzi[1], Erinija Pranckeviciene[1], Arthur R. Summers[1] and Peter Zhilkin[1]

[1] Biomedical Informatics Group, Institute for Biodiagnostics, National Research Council Canada, Winnipeg, MB, Canada.
[2] National Microbiology Laboratory, Winnipeg, MB, Canada.
Corresponding author's e-mail: `Ray.Somorjai@nrc-cnrc.gc.ca`

Summary. While biomedical data acquired from the latest spectroscopic modalities yield important information relevant to many diagnostic or prognostic procedures, they also present significant challenges for analysis, classification and interpretation. These challenges include sample sparsity, high-dimensional feature spaces, and noise/artifact signatures. Since a data-independent 'universal' classifier does not exist, a classification strategy is needed, possessing five key components acting in concert: data visualization, preprocessing, feature space dimensionality reduction, reliable/robust classifier development, and classifier aggregation/fusion. These components, which should be flexible, data-driven, extensible, and computationally efficient, must provide accurate, reliable diagnosis/prognosis with the fewest maximally discriminatory, yet medically interpretable, features.

1 Introduction

Many conventional diagnostic/prognostic clinical procedures are invasive. In addition, their sensitivity and/or specificity are frequently low. The obvious need to replace these with non-invasive or minimally invasive, yet more reliable methodologies has led to the development of spectroscopic, and more recently, microarray-based experimental techniques. Prominent roles are played by *magnetic resonance* (MR), *infrared* (IR), *Raman, fluorescence* and *mass* spectroscopy, all providing spectra of biofluids and tissues. Depending on their frequency range, these methods probe molecular sizes, from diatomic molecules to large proteins. The analysis/classification of these spectra has the goal of distinguishing diseases or disease states. One can also apply the first four methods *in vivo*, a significant advantage in the clinic (consider the possibility and promise of deciding, without surgical intervention, the malignancy of a brain tumor).

Unfortunately, biomedical spectra are plagued by the twin curses of dimensionality and data set sparsity [49]. The first curse is active because the dimensionality

W. Dubitzky and F. Azuaje (eds.), Artificial Intelligence Methods and Tools for Systems Biology, 67–85.

d of spectral feature space is typically very high, $d = O(1\,000)$ to $O(10\,000)$; the spectral features are the various intensity values at the d measurement frequencies. The second curse arises because of the practical difficulty and/or cost of the experimental acquisition of a statistically meaningful number N of biomedical samples; frequently N is only of $O(10)$ to $O(100)$. This leads to a sample to feature ratio (SFR) N/d that is in the $1/20$ to $1/500$ range.

However, the pattern recognition/AI community generally accepts that to create a classifier with high generalization capability, we require an SFR of at least 5, preferably larger [30]. (Cover's combinatorial proof sets the minimum theoretical value of SFR at 2 as the natural separating capacity of a family of d-dimensional surfaces; however, 'the probability of ambiguous generalization is large, unless the number of training patterns exceeds the capacity of the set of separating surfaces' [15], i.e. unless SFR > 2.). This seems to be a necessary condition. However, even if this is satisfied, sufficiency is not guaranteed for small sample sizes; apart from a few recent exceptions [19, 45], this latter caveat is not fully appreciated [49].

It is important to constantly keep in mind the general caveat: 'there are no panaceas in data analysis' [28]. Thus, a best, 'universal' classifier does not exist: the choice of preprocessing methodology, classifier development, etc. is data dependent and should be data-driven. Instead of a futile search for a (nonexistent) 'universal' classifier, we have to formulate and realize a flexible classification **strategy**.

We call our strategy a *statistical classification strategy* (SCS), although it is a hybrid of several statistical and machine learning/AI methods, supporting Breiman's view [13] that statistical and algorithmic modeling are different sides of the same coin and both should be exploited. The SCS evolved over the last decade in response to the need to reliably classify biomedical data that suffer from the above twin curses. In particular, we formulated the strategy with clinical utility in mind: not only should the eventual classifier provide accurate, reliable diagnosis/prognosis, it should also predict class membership using the fewest possible discriminatory features (attributes). Furthermore, these features must be interpretable in biochemically, medically relevant terms ('biomarkers'). These two interrelated aspects are not always appreciated, and thus considered when developing classifiers for biomedical applications.

Because of the twin curses, reliable classification of biomedical data, spectra in particular, is especially difficult, and demands a 'divide and conquer' approach. Thus, our strategy consists of several stages: 1) *Data visualization*, 2) *Preprocessing*, 3) *Feature extraction / selection*, 4) *Classifier development* and 5) *Classifier aggregation / fusion*. We activate all or some of these stages as needed, depending on the data we need to classify. We show the possible interrelations of the various stages in Fig. 1.

In the following, we shall lay out a 'roadmap' of the evolution of the SCS, and describe our decade-long travel along this road. Whenever useful or revealing, we shall also compare the SCS with more standard AI/machine learning approaches, although without pretending to provide a thorough review of the latter. The journey continues, with new challenges and promising future directions; we shall also sketch these.

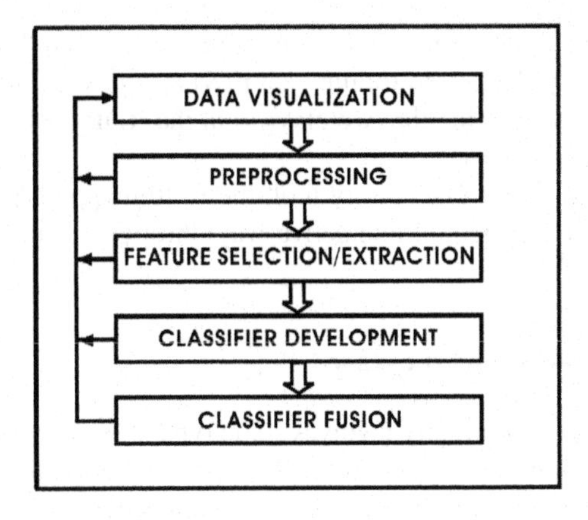

Fig. 1. Flowchart for the statistical classification strategy

2 The Strategy

2.1 Data visualization—a picture is worth a thousand words

Any strategy involving the type of high-dimensional data we want to analyze requires as its essential component, some effective, low-dimensional, yet accurate visualization methodology. Prior to any classifier development, proper data set visualization would help detect 'outliers' (e.g., poor-quality or noisy spectra), and assess whether both training and test sets have been drawn from the same distribution (a necessary condition for a meaningful evaluation of a classifier's generalization power). Such guarantee is particularly important for sparse data sets. We have recently introduced a distance-(similarity)-based, *exact* mapping from $d \gg 2$ to 2 or 3-dimensions [48]. It relies on the simple fact that one can display in two dimensions *exactly* the distances between *any three d-dimensional points* (d arbitrary). We call it mapping to a *relative distance plane* (RDP). It only requires a single distance matrix calculation in some metric (e.g. Euclidean or Mahalanobis). By choosing *any* two points in the feature space (the *reference* pair, $\mathbf{R}_1$, $\mathbf{R}_2$), one preserves the exact distances of all other points to this pair. One can use any pair of points (e.g., the two class centroids), and a line through them defines a potentially 'interesting direction'. Projecting the samples onto this *reference axis* creates a histogram. A multi-peaked histogram indicates data clustering. When projecting the patterns onto some discrete collection of reference axes, we can interpret the RDP mapping as a discretized version of projection pursuit [25]. If classification is of interest, $\mathbf{R}_1$, $\mathbf{R}_2$ should generally come from different classes (but need not be the class centroids). For a 2-class case, with N_1 samples in class 1 and N_2 in class 2, including the two class centroids, there are $(N_1+1)(N_2+1)$ possible reference pairs. We can rank these according to misclassification error; if the latter is zero for several reference pairs, we break ties by ranking them in decreasing order of the training set (TS) *margins* (minimum distances between two patterns separating the two classes) in the RDP [48]. RDP mapping does not require any optimization.

Although primarily an exploratory tool, the RDP mapping is particularly beneficial when used in a dynamic, interactive way, throughout the deployment of our strategy (Fig. 1).

2.2 Preprocessing—setting the stage

Preprocessing of spectra initially involves some type of *normalization* (e.g. 'whitening', or scaling to unit area), *smoothing* (filtering), and/or *peak alignment* (to some external or internal reference). Frequently, we obtain better classification results if we first *transform* the data. Typically, we replace MR spectra by their (numerical) *first derivatives*, or *rank-ordered* variants (*rank ordering* replaces the original features by their ranks. This nonlinear transformation reduces the influence of accidentally large or small feature values.) For IR spectra, *second derivatives* are often beneficial. Other (nonlinear) transformations, applied feature-by-feature, include power, log and exponential operations. All of these preprocessing steps are computationally fast.

We have also introduced two nonlinear data transformations that specifically require parameter optimization [18, 50]. Both approaches improved class separation.

Having prepared the data set, we display it in the RDP, identify and remove potential outliers, and if a split into training—test set is feasible, determine whether the training and test sets appear to come from the same distribution [48]. (If this does not hold, any assessment of the generalization capability of the classifier is suspect.)

Based on our extensive experience, the critical stage of the classification strategy for spectra is the reduction of feature space dimension. Its primary role is to lift the curse of dimensionality, i.e., increase the SFR to the acceptable 5 to 10 range. This is both essential and feasible for spectra, for which the majority of the d spectral features are either redundant (correlated) or irrelevant ('noise').

2.3 Feature space dimensionality reduction

We can accomplish *feature space dimensionality* (FSD) reduction in two, conceptually distinct ways. *Filter* methods do not rely on any knowledge of the eventual classifier algorithm. *Wrapper* methods, which intimately tie FSD reduction to a prediction algorithm, tend to perform better, but are biased and execute slower than the filter methods, especially since they generally use some form of crossvalidation [34].

For either method, FSD reduction comes in two flavors. *Feature selection* produces a subset of the original features. *Feature extraction* is the more general approach: it involves obtaining some functional combination of the original features. For instance, *principal component analysis* (PCA) [29], creates the d new features as different linear combinations (the PCs) of the original d features. However, PCA involves a high degree of feature transformation—the extracted features bear little resemblance to the original ones. This feature scrambling is undesirable for interpretability, e.g., for MR or mass spectra, in which relatively narrow, adjacent spectral regions are frequently signatures of specific chemical compounds.

A conceptually better approach than PCA is to use *partial least squares* for dimensionality reduction. Partial least squares sequentially maximizes the *covariance* between the response variable (the class label) and a linear combination of the original features, hence it is expected that the new features produced will be better class predictors [37]. However, the interpretational difficulties persist because of the PCA-type scrambling.

These conceptual problems led us to develop a specific feature extraction method [38, 39] whose most important interpretational advantage is that it retains spectral identity: the new features are functions (typically the averages) of *adjacent* spectral points.

Before discussing this method in detail, another, more recent approach is worth mentioning. It uses the fact that the PCA basis is not the only orthogonal basis for the subspace spanned by the data points. In fact, one can *rotate* the PCA basis so that the basis functions appear to be, as much as possible, indicator functions supported over an interval of the data. A procedure for determining such a rotation is described in [8]. Using this rotation, one can construct a basis, the *rotated* PC (RPC) basis for the subspace spanned by the data in which the derived features resemble, as much as is possible, averages over a region of the original feature space. When $d > N$, this can be done without loss of information from the original d-dimension data. ¿From these N RPCs one may select maximally discriminatory subsets by exhaustive search.

An alternate method for feature space reduction is to use the discrete wavelet transform with thresholding [42]. The discrete wavelet transform is also an orthogonal transform and converts the original high-dimensional spectra into a set of wavelet coefficients. By thresholding the absolute values of these coefficients, one may drastically reduce the number of extracted features, although the number is not necessarily as low as N. Since the number of thresholded

coefficients varies from sample to sample, the approach taken in [42] was to select those coefficients that satisfy the threshold in $M < N$ of the samples.

Our feature extraction approach is a *genetic-algorithm*-based *optimal region selection* algorithm (GA_ORS) [39]. We chose a genetic algorithm because it is a global minimizer, working on a string of 1s and 0s, a natural representation of the *presence* or *absence* of spectral feature f_k. The inputs of GA_ORS are F, the maximum number of features (distinct subregions) required, and the usual parameters of GA optimization: crossover and mutation rates, number of generations, and number of 'elite' chromosomes. For spectra, n_k consecutive 1s would represent the kth subregion, g_k, k = 1,...,F. GA_ORS minimizes simultaneously the misclassification error and the classification fuzziness (i.e., increases class assignment probabilities). Unique to our method is that we can specify the type of required mathematical operation on adjacent features. We generally use averaging: a new feature g_k is the arithmetic *average* of n_k adjacent original f_ks. Thus, one can view any of the F g_ks as a linear *projection* into an n_k-dimensional subspace, followed by averaging over that subspace. For

MR and mass spectra, such averaging is particularly meaningful, since the averaged regions

frequently estimate specific peak areas. A bonus is that averaging tends to increase the signal-to-noise ratio (i.e., corrects for the presence of noise). We also specify what type of classifier (linear discriminant analysis (LDA), quadratic discriminant analysis, k-nearest-neighbors, etc.) we should use in this wrapper-based feature extraction approach. We usually employ LDA as the classifier, with leave-one-out crossvalidation; using proper feature extraction, even the simplest classifiers frequently outperform their more sophisticated (e.g., nonlinear) counterparts [45].

We incorporated into the GA_ORS software several algorithmic enhancements. Among the most useful are constraints that minimize the maximum misclassification error among the classes (i.e. for 2-class problems equalize false positive and false negative rates), penalize classification errors differently for different classes (relevant when there is considerable imbalance between class sample sizes), and set lower and upper bounds on the allowed spectral subregion widths. (This is important for the mass spectra from proteomics, for which only very narrow regions, 5–20 adjacent mass/charge values, are usually meaningful.)

The GA_ORS approach has been very successful (for its application to the classification of biomedical MR and IR spectra, see the review [36] and references therein). GA_ORS worked equally well for mass spectra from proteomics data [49]. It is even possible to apply the approach to data already reduced using rotated PCA. We recommend the RPC pretreatment because it reduces the dimensionality of the feature space from d ($\gg N$) to N. GA_ORS will then better be able to further reduce the data to a set of features that satisfy the SFR requirement.

For microarrays, the original features are expression levels of *individual* genes, and FSD reduction requires a *feature selection* approach. Since the number of genes is typically $d = O(1\,000)$ to $O(10\,000)$, exhaustive search (ES) (except for single features) is clearly unfeasible, especially for a wrapper-based feature selection method.

A good discussion of the conventional feature selection approaches is in [31]. They claim that the *sequential forward floating selection* (SFFS) method [41] is particularly powerful and near optimal. The SFFS is a generalization of the 'plus-s minus-r' method, which enlarges the feature subset by s features using forward selection, and then deletes r features using backward selection. In SFFS, the values of s and r are determined automatically and updated dynamically. An improved, adaptive version was announced in [47].

Until very recently, feature selection for proteomics and especially for microarray data relied primarily on filter-based approaches, using some univariate criterion for feature ranking, such as in [56]. Typically, the best K (50–200) features were selected for classification, often with $K > N$. With sophisticated classifiers (e.g., neural nets or support vector machines, SVMs) apparently very good results were obtainable. Unfortunately, the reliability of these results is questionable, because of the small N and still large K. Selecting from these K features an optimal subset, although leading to a multivariate set that might even satisfy the SFR requirement, ignores the fact that in general, no nonexhaustive feature selection method can be guaranteed to find the optimal subset. In fact, any ordering of the classification errors obtained with each of the $2^d - 1$ feature subsets is possible [16]. Thus, there is no guarantee that the best *two* or *more* features are amongst the K 'best' features selected via some univariate method.

We found a (necessarily suboptimal) two-stage, ab initio multivariate strategy that produces good classification results.

At the first stage, we use GA_ORS as outlined above, even if no prior grouping of similar genes was carried out (e.g., by an initial clustering). We specify as input $F \ll d$ *temporary* features (averaged subregions) that satisfy the SFR criterion. These F regions comprise n_k adjacent averaged features, $k = 1, \ldots, F$, a total of $d_F = \sum_k n_k$ original features. d_F is generally much less than the original d, typically 1–2 orders of magnitude smaller.

At the second stage, we apply ES to the d_F features for the best (sample-size-dependent) K (2–5) individual features, using a wrapper approach with crossvalidation. We assess the relevance of the individual features by counting their frequency of occurrence in the classifiers tested. This counting is weighted. The weight for feature m is $\sum_j \kappa_j C_j{}^{1/2}$, with $0 \leq C_j \leq 1$ the fraction of crisp ($p \geq 0.75$) class assignment probabilities, and κ_j Cohen's chance-corrected index of agreement [14] for classifier j; the sum is over the number of occurrences of feature m.

This two-stage approach might be viewed as an example of an *embedded* feature selection method [26]. The latter incorporates feature selection as part of the training process. In fact, simultaneous classification and relevant feature identification is an active area of research. For classification of a set $\mathbf{x} = \{\mathbf{x}_i\}$, $i = 1, ..., N$ of d-dimensional vectors, the generic idea is to somehow identify a *sparse* hyperplane, $g(\mathbf{x}) = \mathbf{w} \bullet \mathbf{x} + \mathbf{w}_0$, i.e., automatically eliminate contributions from most of the original d features. Some form of *regularization*, i.e., adding a term such as an L_p norm $||\mathbf{w}||_p$ of the weight vector $\mathbf{w}$, generally induces sparseness. Ideally, one would minimize the L_0 norm $||\mathbf{w}||_0$ of $\mathbf{w}$ forming the hyperplane; in practice, one minimizes the computationally more tractable (weighted) L_1 norm $||\mathbf{w}||_1$. If the data set is not linearly separable, a penalty term, the sum of non-negative slack variables, $\sum_{k=1}^{d} \zeta_k$ with ζ_k given by the hyperplane, $\zeta_k = 1 - y_k\, g(\mathbf{x})$ (an upper bound on the number of misclassifications) alters the objective function $||\mathbf{w}||_1$ to $||\mathbf{w}||_1 + C \sum_{k=1}^{d} \zeta_k$. y_k is the class label of sample k, and is either +1 or –1. The parameter C, $0 \leq C \leq 1$, controls the tradeoff between misclassification and sparseness. The constrained minimization of the objective function can be cast as an efficient linear programming problem, e.g., via LIKNON, as in [6], or linear SVMs, as in [7]. We have compared LIKNON with GA_ORS on several biomedical spectral classification problems. Although there is frequently overlap between the two subsets selected, in general LIKNON produces many more (probably strongly correlated) features than GA_ORS, often more than the number of samples. However, a LIKNON subset could serve as a starting point for additional feature selection.

In Fig. 2 we show the outcomes of applying the GA_ORS and LIKNON feature selection methods to a 2-class biomedical MR spectral data set with 300 original features. There are 61 samples in the healthy class and 79 in the cancer class.

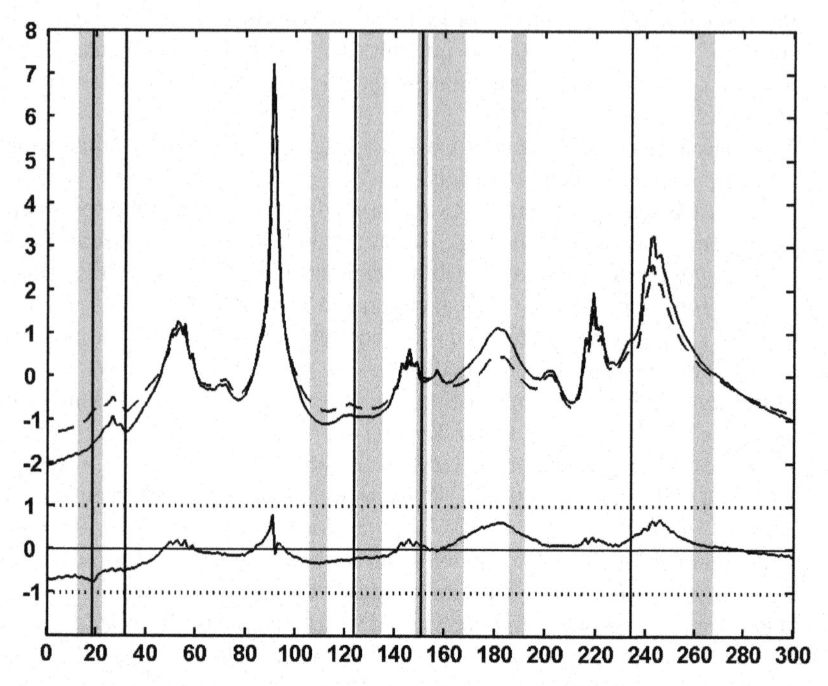

Fig. 2. Results of applying different feature selection methods to a 2-class biomedical MR spectral data set with 300 initial features. We display the *class centroids* (– healthy,– colon cancer), the *difference curve* (healthy – cancer), and both the GA_ORS-created optimal subregions (shaded bars), and the LIKNON-produced single features (black vertical lines).

2.4 Development of reliable classifiers

Having obtained (via GA_ORS or otherwise) F discriminatory features that satisfy the SFR appropriate for the data set size, the next step is to develop a reliable classifier, i.e., one that has high generalization power. Ideally, one would like to partition the data set into a *training set*, a *monitoring* (tuning) set and an independent *validation* (test) set. However, when the sample size is small, such partitioning is not sensible or meaningful. The simplest approach is to use the entire data set when developing the classifier. However, this *resubstitution* method is known to give an optimistically biased error estimate (EE). Amongst the various *crossvalidation* (CV) approaches, the *leave-one-out* (LOO) method uses $N-1$ of the N samples to train on and validates on the left out sample. By leaving out each of the N samples in turn, it provides an essentially unbiased EE, but for small N the variance can be unacceptably large. In $k-fold$ *CV*, we split the data set into k approximately equal parts, train a classifier on $k-1$ parts, validate on the held out portion, and repeat the process, cycling through all k folds. The ultimate EE is the average of k EEs, one for each fold. The *holdout* (i.e., 2-fold CV) method uses approximately half of the data for training and the remainder for validation. Its EE is pessimistically biased. For the k-fold crossvalidation ($k \neq N$, i.e., not LOO), the data can be repeatedly partitioned, preferably in a *stratified* fashion, i.e., maintaining the original sample proportions in the classes—there is evidence that this improves the EE [58]. The *bootstrap* method, a random resampling with replacement strategy [21], helps overcome both large bias and large variance. It produces numerous artificial data sets of the same size as the original; for each, one creates a classifier and averages the classifier outcomes. In the limit, because of replacement, approximately 37.6% of the samples are left out and can be used for validation.

Whatever the error estimation methods, the ultimate classifiers are of the resubstitution type, using the entire data set. Inspired by the 'resampling with replacement' philosophy, we have developed a method to create reliable classifiers [51]. This involves randomly selecting (in a stratified manner) about half of the samples as a training set, developing a crossvalidated classifier, and using the remainder to test the efficacy of the classifier. The training samples are then returned to the original pool and the process repeated, usually $B = 5\,000$ to $10\,000$ times (less if N is very small). We save the optimized classifier coefficients for all B random splits. The *weighted average* of these B sets of coefficients produces the final classifier. The weight for classifier j is $Q_j = \kappa_j C_j^1/2$, with $0 \leq C_j \leq 1$ the fraction of crisp ($p \geq 0.75$) class assignment probabilities, and κ_j is Cohen's chance-corrected measure of agreement [14], $\sim 0 \leq \kappa_j \leq 1$, with $\kappa_j = 1$ signifying perfect classification. The B Q_j values used for the weights are the ones found *not* for the training sets but for the less optimistic *test sets*. We report classifier outcome as *class probabilities*. The weighted Q-average need only include a fraction of the top-ranked classifiers.

2.5 Classifier aggregation/fusion

Classification reliability requires statistically significant (high confidence) class assignments for the samples, i.e. the assignment probabilities p_{mn} should be *crisp*, close to unity (e.g., for 2-class problems p_{mn} is crisp if ≥ 0.75)). If the overall crispness of the single classifier is low, or the accuracy is unsatisfactory, we activate the classifier fusion stage of the SCS. At this stage, one combines the outputs (class probabilities) of several classifiers to form a new classifier. The expectation is that due to aggregation, the accuracy of the combined classifier will increase relative to the accuracy of the best individual classifier.

One assumes that an ensemble of classifiers must be both diverse and accurate for this to happen. However, the accuracy requirement depends on how one combines the individual outcomes.

Diversity ensures that the individual classifiers make independent errors (in addition to the Bayes error); then the error after their aggregation may be reduced by a factor up to the number of individual classifiers combined [55]. One of the important theoretical results of classifier aggregation is the ambiguity decomposition [35]. It states that the mean-square error of the combined error estimate is guaranteed to be less than or equal to the average mean-square error of the component error estimates.

An obvious (not necessarily optimal) way to promote diversity is to apply several conceptually and methodologically diverse prediction methods to the data set. We used this approach with success to discriminate between the MR spectra of cancerous and normal thyroid neoplasms [52]. LDA, a neural net-based method and genetic programming formed the triplet of classifiers, applied to two different spectral regions. The median of the six probabilities produced was the combiner. 100% sensitivity and 98% specificity was obtained for an independent validation set.

Theoretical considerations notwithstanding, in our experience, classifier diversity does not seem to be critical. Thus, classifier fusion need not rely on widely different classifier methodologies. We found that for classifying biomedical spectra, using a single, simple classifier, e.g. LDA, preprocessing the spectra differently (taking different derivatives, rank ordering, etc) and/or selecting different spectral regions will provide sufficient diversity for classifier aggregation.

Since the mid 90s, there has been an explosion of research on classifier fusion (under many other appellations). Conceptually important approaches of practical interest include *bagging* [9], *boosting* [23], *output randomization* [12], *arcing* [10], and variants [5, 54]. There are annual conferences and workshops devoted to 'Multiple Classifier Systems', see e.g., [57] and earlier proceedings. We strongly advise consulting these.

Let us suppose that K different classifiers are available (for simplicity, we assume 2-class classification), and we want to combine their outcomes (a useful analogy: there are K experts with K possibly different opinions; we want to reach an optimal consensus). For any sample $\mathbf{x}$, the K classifiers produce K class probabilities $p_k(\mathbf{x})$. The following is one of numerous possible categorizations of approaches that combine multiple classifier outcomes. (The literature is extensive.)

1) 'Passive': These approaches are data-independent. The AI community typically uses the averages or medians or even products of the $p_k(\mathbf{x})$s [53]. We shall mention other combining methods below.

2) 'Active': The prime example of this is the multi-classifier variant of *stacked generalization* (SG) [59]. It uses the K $p_k(\mathbf{x})$s (level 0 outputs of the first set of K classifiers) as K *inputs* to another (level 1) classifier. This process can be repeated (levels 2, 3,...).

We have compared several aggregation techniques on various artificial and real-life MR and IR data in [60]: *Averaging, majority voting, logistic regression, linear combination, fuzzy integration, entropy, confidence factor*, and *stacked generalization*. Most of the techniques require training, usually involving the minimization of an objective function. Training is computationally fast for linear combination and quite time-consuming for fuzzy integration and logistic regression. The entropy and confidence factor approaches do not require any training. The level 1 classifier we used for stacked generalization was the relatively fast LDA/LOOCV.

Based on these results, our general conclusion (confirmed many times since) is that for biomedical spectra:

- Aggregating classifiers will in most cases lead to a better classification performance than that of the best individual classifier, and almost invariably results in more crisp classification. This latter observation is particularly relevant in a clinical setting, where diagnosis (classification) with high confidence is important.

- Preprocessing data differently and submitting these to a single, reliable (level 0) classifier is a simple and viable technique for creating diversity.
- Just as it is difficult to choose the best among different classifiers, it is also difficult to choose the best aggregation method. Different methods perform well on some data and poorly on others. For biomedical data, SG with a simple level 1 classifier (LDA/LOO) generally gives high classification accuracies and produces crisp (high class assignment probability) outcomes.

For a reliable and realistic error estimate, especially for small N, it is important to use crossvalidation *at all stages*: at feature selection/extraction, at classifier development and even at classifier fusion [4, 45].

3 Challenges and Some Partial Solutions—Through a Glass Darkly

3.1 Non-uniqueness

Non-uniqueness is largely due to small sample size: several sets of features give comparable classification accuracies (sensitivities, specificities), not only for the training set but also for the independent validation set [32, 33, 49]. Given equally good feature sets, we found that using the RDP mapping can help in deciding which will generalize better. We demonstrate this on the prostate cancer mass spectroscopy data set 'JNCI 7-3-02', from (http://clinicalproteomics.steem.com). 'JNCI 7-3-02' has 42 samples in class 1 (black disks) and 42 in class 2 (white disks) in the training set (TS), and 21 class 1 (black triangles) and 27 class 2 (white triangles) samples in the validation set (VS) (Fig. 3). There are 15 154 original 'features' (mass/charge, M/Z values), giving an SFR per class of $42/15\,154 \approx 1/361$, instead of the recommended 5 to 10.

In the left panel of Fig. 3 we show results of the mapping (in the L_2 norm) from the 15 154-dimensional feature space to the RDP. This produced 8 misclassifications in the TS and 9 in the VS, likely the worst possible result with a linear classifier.

Using an LDA/LOO-wrapper-based feature selection from the original 15 154-dimensional feature space (1 million random sets with replacement), two different 5-dimensional feature sets (set A and set B), the features being the intensity values at five M/Z positions, had no misclassification error for either TS or VS [49]. Which of these will generalize better? On mapping from 5 dimensions to the RDP, this classification-based ambiguity disappears: the classifier with set B is more likely to generalize better than the classifier with set A; in the L_2 norm, set A gave 7 misclassifications in the TS and 2 in the VS (middle panel), whereas set B produced 8 reference pairs with zero misclassification for the TS, three of which (one displayed in the rightmost panel) also produced no error in the VS. If we use the Mahalanobis distance as the metric for the RDP mapping, there are many more reference pairs

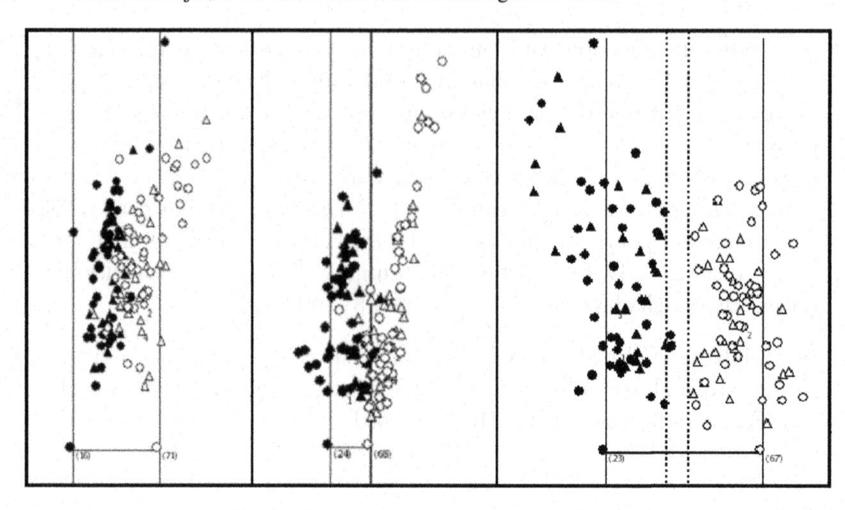

Fig. 3. Mapping to the RDP (L_2 norm). Left panel: From 15 154 features (M/Z values). 8 (TS), 9 (VS) misclassifications. Middle panel: From 5 dimensions (feature set A). 7 (TS), 2 (VS) misclassifications. Right panel: From 5 dimensions (feature set B). 0 (TS), 0 (VS) misclassifications.

with perfect TS + VS classification. (This is to be expected, since feature selection was LDA-based.) However, a classifier using feature set B still appears superior: 174 of the 1 849 possible reference pairs give perfect TS + VS results, whereas only 12 of the 1 849 pairs are error-free for feature set A.

3.2 Open vs. closed system

Given a 2-class classifier C_{mn}, it will necessarily assign a new exemplar $\mathbf{x}$ to either class m or to class n, i.e., C_{mn} presents a *closed* classifier system. In contrast, an *open* system allows for a *reject class*: exemplars that do not appear to belong to either class should be identifiable. This is a generic problem and is closely tied to outlier detection, with all its attendant difficulties. The *atypicality index* [2, 3] of $\mathbf{x}$ attempts to assess how different $\mathbf{x}$ is from the population it is compared to. An exemplar $\mathbf{x}$ may be placed in the reject class because it is too far from either class ('distance reject') or because its class assignment is undecidable ('ambiguity reject') [20]. It is somewhat simpler to ascertain the latter condition if C_{mn} assigns class probabilities; we call the class assignment of $\mathbf{x}$ *ambiguous* or *fuzzy*
if $0.25 \leq p_{mn}(\mathbf{x}) \leq 0.75$.

3.3 The K-class problem—it is better with pairs

Proponents of multi-class methods argue that the larger the number of classes, the less likely that a 'random' set of features provides good discrimination. However, in the case of uneven distributions across classes, multi-class methods over-represent abundant or easily separable classes [22]. Extensive experimentation suggests that for the problems we considered, the best K-class classifier is obtained when we develop $K(K-1)/2$ 2-class classifiers and combine their outcomes. When $K(K-1)/2$ is large, an alternative approach is to create K 2-class classifiers, class J vs. the $K-1$ remaining classes, $J = 1, \ldots, K$, [43], or in general, use error-correcting output codes [17]. Care must be taken to deal with the inevitable sample size imbalance. A simple remedy we found effective is to assign (by crossvalidation or trial-and-error) a larger weight to the less populous class when developing the classifier.

Two-class classifiers have several advantages. They provide a more flexible model than a K-class classifier (e.g., LDA assumes that all K classes have the same covariance matrix. This would be much more restrictive than assuming a common covariance matrix for only two classes: we need to pool the K potentially different covariance matrices only two at a time). Furthermore, each 2-class classifier may assume a different optimal feature set, thus providing additional classification flexibility. In addition, unlike for K classes, the reliable detection of outliers in the 2-class case is much less problematic. Finally, for a 2-class problem, a linear classifier and a linear regressor are equivalent. However, being able to treat classification as regression, robust versions, e.g. *least trimmed squares* [1] can be developed more naturally.

3.4 Methods of combining 2-class classifiers ('decoding')

Let $p_{mn}(\mathbf{x})$ be the probability that when a d-attribute sample $\mathbf{x} = (x_1, \ldots, x_d)$ is submitted to C_{mn}, the pair classifier for classes m and n, it will be assigned to class m. To compute the *posteriori probabilities* $p_m(\mathbf{x})$ for all $m = 1, 2, \ldots, K$, we present $\mathbf{x}$ to all C_{mn}, and obtain the corresponding probabilities $p_{mn}(\mathbf{x})$. Given the $p_{mn}(\mathbf{x})$, Friedman [24] introduced a simple combination rule: Assign $\mathbf{x}$ to the class that wins the most pairwise comparisons. This is an integer count of all pair probabilities $p_{mn}(\mathbf{x}) \geq 0.5$. A natural extension is to use the actual probability values $p_{mn}(\mathbf{x})$. Then

$$p_m^{F^*}(\mathbf{x}) = (2/K(K-1)) \sum_{n \neq m} p_{mn}(\mathbf{x}) \quad (1)$$

$$R_{F^*} = \arg\max_m [p_m^{F^*}(\mathbf{x})] \quad (2)$$

$p_m^{F^*}(\mathbf{x})$ is now a proper (unnormalized) probability. In fact, Eq. 1 provides the starting values for an iterative, maximum-likelihood method introduced in [27].

An alternate approach to these 'winner takes all' rules has been proposed in [44]. It is based on an optimal 'network' design of the K-class classifier, derived from all possible 2-class classifiers, and leads to the combining rule

$$p_m^B(\mathbf{x}) = 1/[1 + \sum_{n \neq m} (1/p_{mn}(\mathbf{x}) - 1) \quad (3)$$

Eq. (3) is the critical relation that connects the *a posteriori* probability $p_m(\mathbf{x})$ that $\mathbf{x}$ belongs to class m to the *a posteriori* probabilities $p_{mn}(\mathbf{x})$ obtained from the $K-1$ pair classifiers trained to discriminate class m from the $K-1$ classes, $n = 1, 2, \ldots K, n \neq m$. The K $p_m^B(\mathbf{x})$s have to be normalized by dividing each by their sum. In our experience, Eq. (3) generally outperformed both Friedman's and Hastie's [27] approach.

We generalized Eq. (3) to account for the *quality* of the 2-class classifiers C_{mn}. This leads to quality-weighted final probability assignments. If $Q_{mn} = Q_{nm}(0 \leq Q_{mn} \leq 1, \sum_{n \neq m} Q_{mn} = 1)$ is some normalized measure of the quality of classifier C_{mn}, then Eq. (3) becomes

$$p_m^B(\mathbf{x}) = 1/[\sum_{n \neq m} (Q_{mn}/p_{mn}(\mathbf{x}))] \tag{4}$$

3.5 Burnishing tarnished gold standards

The problem of extracting diagnostic information is often exacerbated by the fact that the 'gold standard'—the external reference test against which a newly proposed and possibly imperfect diagnostic test is measured—may itself be imprecise or even unreliable. Factors contributing to a 'tarnished' gold standard include subjective estimates by a domain expert (or panel of experts), simple clerical errors, unreliable or imperfect sample acquisition techniques, or anomalous sensor readings. However, little work has been done to investigate a methodology whereby the possible imprecision of a well-established gold standard may be addressed while at the same time maintaining its essential discriminatory power. One effective strategy for dealing with this problem is the gold standard adjustment of spectra in a training set using a robust estimation of deviations from class medians [40].

The strategy begins with calculating the medoid (a robust centroid) for each class. One then uses a multivariate extension of the median of absolute deviations (robust to outliers and long-tailed distributions) to compute the robust estimates of dispersion for each class medoid. We apply a weighted distance measure to each training set spectrum for each class medoid. We incorporate this distance measure into the original gold standard, using a fuzzy set theoretic membership function. While the original gold standard assigns each spectrum to one and only one class, the fuzzy gold standard adjustment assigns each spectrum to all classes, but to varying degrees. If a spectrum is near its class medoid, the original gold standard predominates. If it is far from its class medoid, its membership in its original class is reduced. Moreover, if it is also sufficiently near another class medoid, its membership in that class will increase. Note that if a class label adjustment occurs, it only occurs for spectra within the training set; test sets are never altered.

This method was applied to a set of 206 normalized ^{1}H MR spectra of human brain neoplasms (95, 74, and 37 spectra in three classes) that were randomly assigned to training and test sets (80 and 126 spectra, respectively). The method adjusted significantly class labels for 3 training spectra; this improved the overall test set accuracy of the classifier (a multi-layer perceptron) to 87% (versus 83% with the original gold standard). If we simply remove the three suspect spectra from the training set, as is often done with 'outliers', the test set accuracy drops to 77%.

3.6 Nonlinear transformations

Most classes are not separable linearly. To handle arbitrary decision boundaries, we either have to use some nonlinear classifier (e.g., neural nets) in the original feature space, or first we have to transform the features. The latter approach is more attractive because the appropriate transformation may render the classes *linearly separable* in the *transformed space*. Such linearization carries penalties with it. The number of parameters (i.e., the classifier coefficients) to be optimized will increase. This is undesirable for biomedical data sets both because of interpretational difficulties and because they already suffer from sparsity.

The kernel 'trick' of SVMs, i.e., using a nonlinear mapping of the original features into an even higher dimensional feature space, may give an apparently error-free classification (very high-dimensional spaces are essentially empty), but at the expense of interpretability. On the other hand, after appropriate feature space reduction (to a low enough dimension to also satisfy the SFR), a nonlinear mapping converts nonlinear class boundaries in the reduced space into hyperplanes in the new feature space. The dimensionality of this space will still be low enough to be interpretable, and one can exploit the simplicity and robustness of linear classifiers. Such considerations further strengthen the argument for feature space reduction.

4 Future Directions and Persisting Challenges

The increased and increasing emphasis in the biomedical arena on certain data acquisition methods (microarrays and various forms of spectroscopy) will lead to very high-dimensional data sets; the clinical reality is that these data sets are sparse. From the viewpoint of a data analyst/statistician, these twin curses pose a serious challenge. There has been some progress. There are now effective algorithms and strategies to find the *intrinsic* dimensionality of high-dimensional feature spaces: we can deal with the curse of dimensionality. The curse of data set sparsity is another matter: it is often not cost-effective, feasible or practical to experimentally increase sample size. So what can one do to create diagnostic/prognostic tools ('classifiers') that are reliable?

A possible answer is to devise methods that generate representative *surrogate* (pseudo) *data*. The hope is that these data will somehow 'fill the gap' in the very sparsely sampled populations. We believe this is a promising direction of research, still in its infancy. In particular, there are already attempts in the literature, if somewhat unsatisfactory. Among these are the convex pseudo-data generation method [11], K-nearest-neighbors-directed noise injection [46] and a 'sample neighborhood smearing' method [32]. We have also started experimenting with a permutation-based approach, with some success. We shall report on these experiments elsewhere.

Clinical reality forces us to accept that not all class label assignments are equally certain. Classification/regression methods that can incorporate such uncertainties will gain more prominence. We have already mentioned one data-driven approach to deal with tarnished gold standards. We are exploring, for 2-class problems, a regression formulation, with domain expert confidence built into the class labels.

5 Acknowledgments

We thank all our collaborators who provided the biomedical spectra, and the owners of the microarray and proteomics data sets for making them publicly available.

References

1. J. Agulló. New algorithms for computing the least trimmed square regression estimator. *Computational Statistics & Data Analysis*, 36:425–439, 2001.
2. J. Aitchison and I.R. Dunsmore. *Statistical Prediction Analysis*. Cambridge University Press, Cambridge, UK, 1975.
3. J. Aitchison, J.D.F. Habbema, and J.W. Kay. A critical comparison of two methods of statistical discrimination. *Appl Statist*, 26:15–25, 1977.
4. C. Ambroise and G.J. McLachlan. Selection bias in gene extraction on the basis of microarray gene-expression data. *PNAS*, 99:6562–6566, 2002.
5. E. Bauer and R. Kohavi. An empirical comparison of voting classification algorithms: Bagging, boosting, and variants. *Machine Learning*, 36:105–139, 1999.
6. C. Bhattacharyya, L.R. Grate, A. Rizki, D. Radisky, and F.J. Molina et al. Simultaneous classification and relevant feature identification in high-dimensional spaces: application to molecular profiling data. *Signal Processing*, 83:729–743, 2003.
7. J. Bi, K.P. Bennett, M. Embrechts, C.M. Breneman, and M. Song. Dimensionality reduction via sparse support vector machines. *J Machine Learning Research*, 3:1229–1243, 2003.
8. C. Bowman, R. Baumgartner, and R.L. Somorjai. Rotated principal component analysis to preserve spatial structure. *to be submitted*, 2004.
9. L. Breiman. Bagging predictors. *Machine Learning*, 26:123–140, 1996.
10. L. Breiman. Arcing classifiers. *Annals of Statistics*, 26:801–849, 1998.
11. L. Breiman. Using convex pseudo-data to increase prediction accuracy. Technical report, U.C. Berkeley, Department of Statistics, 1998. Retrieved from: http://www.biostat.harvard.edu/courses/individual/bio229/lectures/L8/pseudodata.pdf.
12. L. Breiman. Randomizing outputs to increase prediction accuracy. *Machine Learning*, 40:229–342, 2000.
13. L. Breiman. Statistical modeling: The two cultures. *Statistical Science*, 16:199–231, 2001.
14. J. Cohen. Weighted kappa: Nominal scale agreement with provision for scaled disagreement or partial credit. *Psychological Bulletin*, 70:213–220, 1968.
15. T.M. Cover. Geometrical and statistical properties of systems of linear inequalities with applications in pattern recognition. *IEEE Transactions on Electronic Computers*, EC-14:326–334, 1965.

16. T.M. Cover and J.M. van Campenhout. On the possible orderings in the measurement selection problem. *IEEE Transactions on Systems, Man and Cybernetics*, pages 657–661, 1977.
17. T.G. Dietterich and G. Bakiri. Solving multiclass learning problems via error-correcting output codes. *J Artificial Intelligence Research*, 2:263–286, 1995.
18. B. Dolenko and R.L. Somorjai. Time well spent: preprocessing of mr spectra for greater classification accuracy. In *Proceedings of the Society of Magnetic Resonance : third scientific meeting and exhibition and the European Society for Magnetic Resonance in Medicine and Biology, Twelfth Annual Meeting and Exhibition held jointly Nice Acropolis, Nice, France, August 19-25, 1995*, Berkeley, CA, USA, 1995. The Society of Magnetic Resonance.
19. E.R. Dougherty. Small sample issues for microarray-based classification. *Comparative and Functional Genomics*, 2:28–34, 2001.
20. B. Dubuisson and M. Masson. A statistical decision rule with incomplete knowledge about classes. *Pattern Recognition*, 26:155–165, 1993.
21. B. Efron and R.J. Tibshirani. *An Introduction to the Bootstrap*. Chapman Hill, New York, NY, 1993.
22. G. Forman. An extensive empirical study of feature selection metrics for text classification. *J Machine Learning Research*, 3:1289–1305, 2003.
23. Y. Freund and R.E. Schapire. Experiments with a new boosting algorithm. In L. Saitta, editor, *Machine learning : proceedings of the thirteenth international conference (ICML '96), Bari, Italy, July 3-6, 1996*, San Francisco, CA, USA, 1996. Morgan Kaufmann.
24. J. Friedman. Another approach to polychotomous classification. Technical report, Stanford University, 1996.
25. J. Friedman and J.W. Tukey. A projection pursuit algorithm for exploratory data analysis. *IEEE Transactions on Computing*, pages 881–889, 1974.
26. I. Guyon and A. Elisseeff. An introduction to variable and feature selection. *J Machine Learning Research*, 3:1157–1182, 2003.
27. T. Hastie and R. Tibshirani. Classification by pairwise coupling. *Annals of Statistics*, 26:451–471, 1998.
28. P.J. Huber. Projection pursuit. *Annals of Statistics*, 13:435–475, 1985.
29. J.E. Jackson. *A User's Guide to Principal Components*. Wiley, New York, NY, 1991.
30. A.K. Jain and B. Chandrasekaran. Dimensionality and sample size considerations in pattern recognition practice. In P.R. Krishnaiah and L.N. Kanal, editors, *Handbook of Statistics*, volume 2, pages 835–855. North Holland Publ. Co., Amsterdam, 1982.
31. A.K. Jain, R.P.W. Duin, and J. Mao. Statistical pattern recognition: A review. *IEEE Transactions on Pattern Analysis and Machine Intelligence*, 22:4–37, 2000.
32. S. Kim, E.R. Dougherty, J. Barrera, Y. Chen, and M.L. Bittner et al. Strong feature sets from small samples. *J Computational Biology*, 9:127–146, 2002.
33. S. Kim, E.R. Dougherty, I. Shmulevich, K.R. Hess, and S.R. Hamilton et al. Identification of combination gene sets for glioma classification. *Molecular Cancer Therapeutics*, 1:1229–1236, 2002.
34. R. Kohavi and G.H. John. Wrappers for feature subset selection. *Artificial Intelligence*, 97:273–324, 1997.
35. A. Krogh and J. Vedelsby. Neural network ensembles, cross validation, and active learning. *Advances in Neural Information Processing Systems*, 7:231–238, 1995.
36. C.L. Lean, R.L. Romorjai, I.C.P. Smith, P. Russell, and C.E. Mountford. Accurate diagnosis and prognosis of human cancers by proton mrs and a three-stage classification strategy. *Annual Reports on NMR Spectroscopy*, 48:71–111, 2002.

37. D.V. Nguyen and D.M. Rocke. Tumor classification by partial least squares using microarray gene expression data. *Bioinformatics*, 18:39–50, 2002.
38. A. Nikulin, K.M. Brière, L. Friesen, I.C.P. Smith, and R.L. Somorjai. Genetic algorithm-guided optimal attribute selection: a novel preprocessor for classifying mr spectra. In *Proceedings of the Society of Magnetic Resonance : third scientific meeting and exhibition and the European Society for Magnetic Resonance in Medicine and Biology, Twelfth Annual Meeting and Exhibition held jointly Nice Acropolis, Nice, France, August 19-25, 1995*, Berkeley, CA, USA, 1995. The Society of Magnetic Resonance.
39. A.E. Nikulin, B. Dolenko, T. Bezabeh, and R.L. Somorjai. Near-optimal region selection for feature space reduction: novel preprocessing methods for classifying mr spectra. *NMR Biomed*, 11:209–216, 1998.
40. N.J. Pizzi. Fuzzy pre-processing of gold standards as applied to biomedical spectra classification. *Artificial Intelligence in Medicine*, 16:171–182, 1999.
41. P. Pudil, J. Novovicová, and J. Kittler. Floating search methods in feature selection. *Pattern Recognition Letters*, 15:1119–1125, 1994.
42. Y. Qu, B.-L. Adam, M. Thornquist, J.D. Potter, and M.L. Thompson et al. Data reduction using discrete wavelet transform in discriminant analysis of very high dimensionality data. *Biometrics*, 59:143–151, 2003.
43. R. Rifkin and A. Klautau. In defense of one-vs-all classification. *Journal of Machine Learning Research*, 5:101–141, 2004.
44. J. Schürmann. *Pattern Classification - a Unified View of Statistical and Neural Approaches*. John Wiley & Sons, Inc, New York, NY, 1996.
45. R. Simon, M.D. Radmacher, K. Dobbin, and L.M. McShane. Pitfalls in the use of dna microarray data for diagnostic and prognostic classification. *Journal of the National Cancer Institute*, 95:14–18, 2003.
46. M. Skurichina, S. Raudys, and R.P.W. Duin. K-nearest neighbors directed noise injection in multilayer perceptron training. *IEEE Transactions on Neural Networks*, 11:504–511, 2000.
47. P. Somol, P. Pudil, J. Novovicová, and P. Paclík. Adaptive floating search methods in feature selection. *Pattern Recognition Letters*, 20:1157–1163, 1999.
48. R.L. Somorjai, A. Demko, M. Mandelzweig, B. Dolenko, and A.E. Nikulin et al. Mapping high-dimensional data onto a relative distance plane - an exact method for visualizing and characterizing high-dimensional patterns. *Computational Statistics & Data Analysis*, submitted, 2004.
49. R.L. Somorjai, B. Dolenko, and R. Baumgartner. Class prediction and discovery using gene microarray and proteomics mass spectroscopy data: curses, caveats, cautions. *Bioinformatics*, 19:1484–1491, 2003.
50. R.L. Somorjai, B. Dolenko, W. Halliday, D. Fowler, and N. Hill et al. Accurate discrimination between low- and high-grade human brain astrocytomas: robust multivariate analysis of 1h mr spectra. *J Medicine and Biochemistry*, 3:17–24, 1999.
51. R.L. Somorjai, B. Dolenko, A. Nikulin, P. Nickerson, and D. Rush et al. Distinguishing normal from rejecting renal allografts: application of a three-stage classification strategy to mr and ir spectra of urine. *Vibrational Spectroscopy*, 28:97–102, 2002.
52. R.L. Somorjai, A.E. Nikulin, N. Pizzi, D. Jackson, and G. Scarth et al. Computerized consensus diagnosis: a classification strategy for the robust analysis of mr spectra. i. application to 1h spectra of thyroid neoplasms. *Magn Reson Med*, 33:257–263, 1995.
53. D.M.J. Tax, M. van Breukelen, R.P.W. Duin, and J. Kittler. Combining multiple classifiers by averaging or by multiplying? *Pattern Recognition*, 33:1475–1485, 2000.

54. A. Tsymbal, S. Puuronen, and I. Skrypnyk. Ensemble feature selection with dynamic integration of classifiers. In *Int ICSC Congress on Computational Intelligence Methods and Applications CIMA'2001, Bangor, Wales, U.K.*, Canada, 2001. ICSC.
55. K. Tumer and J. Ghosh. Error correlation and error reduction in ensemble classifiers. *Connection Science*, 8:385–404, 1996.
56. V.G. Tusher, R. Tibshirani, and G. Chu. Significance analysis of microarrays applied to the ionizing radiation response. *PNAS*, 98:5116–5121, 2001.
57. T. Windeatt and F. Roli. *Multiple Classifier Systems*. Springer, Berlin, New York, 2003.
58. I.H. Witten and E. Frank. *Data Mining*. Academic Press, San Diego, CA, 2000.
59. D.H. Wolpert. Stacked generalization. *Neural Networks*, 5:241–259, 1992.
60. P.A. Zhilkin and R.L. Somorjai. Application of several methods of classification fusion to magnetic resonance spectra. *Connection Science*, 8:427–442, 1996.

Cooperative Metaheuristics for Exploring Proteomic Data

Robin Gras[1], David Hernandez[1], Patricia Hernandez[1], Nadine Zangger[1], Yoan Mescam[12], Julien Frey[1] Olivier Martin[1], Jacques Nicolas[2] and Ron D. Appel[13]

[1] Swiss Institute of Bioinformatics, Geneva, Switzerland.
[2] IRISA-INRIA, Rennes, France.
[3] University of Geneva, Geneva, Switzerland.
Corresponding author's e-mail: robin.gras@isb-sib.ch

Summary. Most combinatorial optimization problems cannot be solved exactly. A class of methods, called metaheuristics, has proved its efficiency to give good approximated solutions in a reasonable time. Cooperative metaheuristics are a sub-set of metaheuristics, which implies a parallel exploration of the search space by several entities with information exchange between them. Several improvements in the field of metaheuristics are given. A hierarchical approach resting on multiple levels of cooperative metaheuristics is presented. Some applications of these concepts to difficult proteomics problems, including automatic protein identification, biological motif discovery and multiple sequence alignment are presented. For each application, an innovative method based on the cooperation concept is given and compared with classical approaches.

1 Introduction

An important challenge of system biology is to analyze the huge amount of data generate by the new field of molecular biology: *proteomics*. Proteomics [33][42] can be defined as the study of the protein expression pattern of a given tissue or organism at a given time. This involves knowing about large number of different proteins, their possible variants (modifications, mutations, fragments ...), their corresponding amino acid sequence and potential interactions between these proteins.

The commonly used technique for proteome analysis involves four steps (see Fig. 1): protein separation (for example by two-dimensional electrophoresis), protein digestion which produces a set of peptides, measurement of the peptides and peptide fragments masses by mass spectrometry, comparison of mass data with proteic or translated genomic sequence databases. Matched masses are used to identify proteins and their possible variants. The understanding of the possible biological functions of the identified new proteins, protein variants or protein complexes rely on various information sources (experimental data, sequence and/or annotation databases, literature...). This approach must be automated or partially automated to analyze in real

W. Dubitzky and F. Azuaje (eds.), Artificial Intelligence Methods and Tools for Systems Biology, 87–106.

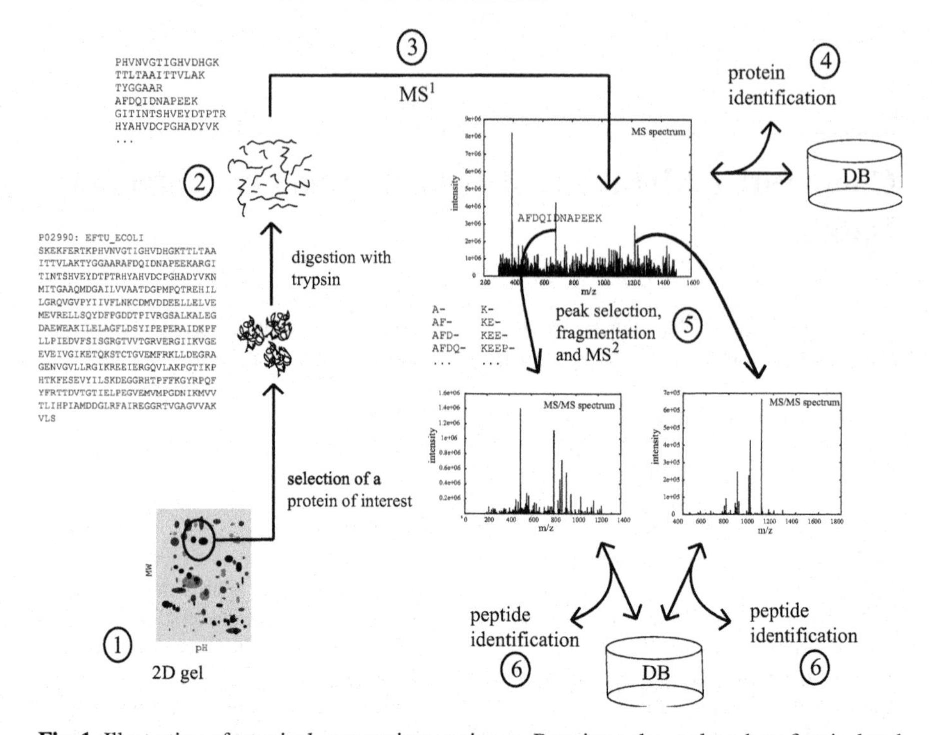

Fig. 1. Illustration of a typical proteomic experiment. Proteins to be analyzed are first isolated from the sample by any separation technique, as for example two-dimensional electrophoresis (1). A protein of interest is selected from the gel and is degraded enzymatically into smaller pieces, the peptides (2). After ionization, the peptides are analyzed by mass spectrometry (MS1) (3), leading to an MS spectrum containing a list of mass-charge ratios (m/z) together with their intensities (denoted as peaks). The source protein of an MS spectrum can be identified by comparing the experimental peak list with virtual MS spectra computed from theoretical proteins stored in databases (4). Additional information can be obtained using tandem mass spectrometry (MS/MS, or MS2). In this approach, a particular m/z value is isolated from the MS spectrum and further fragmented by collision with gas molecules giving rise to fragment ions. Again, the mass-charge ratios of the fragment ions are measured by mass spectrometry (5), and an MS/MS spectrum is produced. While MS spectra correspond to a protein, MS/MS spectra are associated with a peptide. MS/MS spectra can be identified in the same way as MS spectra by computing virtual MS/MS spectra from theoretical peptides and measuring their similarity (6). Alternatively, the sequence of the source peptide can be inferred de novo without information from a database by analyzing the relative positions of the peaks.

time the large amount of data generated in high-throughput environments commonly used in recent years.

In our research group, we develop different methods for the automation of proteomics data analysis. We are particularly interested in the problems of automatic proteins identification and protein sequences analysis. From our point of view, most of these analyzes can be regarded as difficult combinatorial optimization problems such as, for example, efficient learning of properties from data, classification of complex sets of information, extraction of grammatical structure from sequences, etc. We apply *cooperative metaheuristic* approaches in a hierarchical way to manage a wide variety of proteomics problems.

Metaheuristics are generic methods for non-exact solving of difficult ($NP-hard$ that is problems for which there is no proof that they can be solved in a polynomial time) combinatorial problems [27]. Their global strategy consists of an efficient exploration of the *search space* in order to localize reasonably 'good' solutions for a given objective function. A large variety of such optimization methods is available (*simulated annealing, branch and bound, tabu search, evolutionary algorithms ...*); they can be classified according to different criteria reflecting a particular property to be emphasized. For example, they are often classified into deterministic and non-deterministic categories depending on the use (or not) of a stochastic process for the exploration of the search space.

We are interested in another metaheuristics property leading to classify between cooperative and non-cooperative methods. The non-cooperative metaheuristics are those which explore a unique point of the search space at a given time, like hill climbing, simulated annealing, taboo search, etc. Cooperative metaheuristics correspond to a parallel exploration of the search space by a set of coexisting potential solutions; each solution cooperate with the others by information exchange in order to select new promising potential solutions. This aspect is strongly linked with the concept of *building blocks* (BB) [11, 12], that are relevant sub-parts of solutions shared by most of the good solutions, because information exchange is a way to detect and transmit building blocks between solutions. We distinguish three classes of cooperative metaheuristics according to the origin of the decision and the nature of information exchange. In the first one, referred to as 'centralized cooperation', cooperation between entities is handle by an external oracle that selects both the cooperating entities and the content of the exchange. For example, in a classical evolutionary algorithm, cooperation consists of sub-solution exchange performed by the crossover operator. The second class, referred to as 'individual cooperation', corresponds to a system in which entities manage communication themselves however without any prior information from the other entities. For example, the parallel version of evolutionary programming, based on the *injection island model* [9, 13, 24], is a two level cooperation metaheuristics.

The first level is the centralized cooperation between chromosomes on each island and the second level is an individual cooperation between islands. In this second level, each island sends data to its own selection of other islands independently of the other agents needs. The island model not only provides efficient exploration ability but it is also a common way to preserve diversity [35]. Finally, the third class, referred to as 'concerted cooperation', embodies a cooperation process in which each exchange depends on a mutual agreement between entities. Therefore, the direction and the content of the communication are defined dynamically, in order to optimize the gain of each participant. This model can be much more complex than those described in the other two classes. It implies the definition of a strategy of cooperation for each entity based on knowledge regarding its proper state and possibility of improvement, the information content of the other entities, the strategies of the other entities, etc. However, the exploration of the search space is much more dynamic and 'intelligent' due to the association of two factors: first, the diversity, which is preserved by isolation of potential solutions inside agents; second, the flexible and directed property of the information exchange mechanism between entities.

2 Genetic programming applied to protein identification

Protein identification is a key issue in proteomics. In a global approach, the challenge is to identify all proteins present in a sample. In high-throughput identification projects, the identification tool should be fast, fully automated and robust. Alternatively, proteins of clinical interest can be targeted by differential expression between two samples. In this case, the identification tool must be generic enough to be able to identify mutated or modified proteins.

Nowadays, the most widely used technique in protein identification is mass spectrometry (MS). After purification, each protein is digested using a specific enzyme. The masses of the resulting peptides are then measured. The obtained mass list, called a MS spectrum, may already be used for identification by 'peptide mass fingerprint', but additional information on the protein sequence can improve the accuracy of identification: each peptide is further fragmented. The fragmentation ideally occurs on the peptidic bonds joining together the amino acids and generates ionic fragments carrying one or several charges. The fragment masses are measured, leading to a MS/MS spectrum, which includes the molecular weight of the unfragmented peptide (*the source peptide*) and a peak list representing the masses and the intensities of the detected ionic fragments. The identification relies on a scoring function that compares the experimental MS/MS spectrum with peptides from a database (*theoretical peptide*).

MS/MS identification is not easy, since the fragmentation process is hardly foreseeable and depends, among other things, on the amount of energy used by the mass spectrometer, on the number and the repartition of the charges carried by the peptide, and on its sequence. As a result, some positions on the peptide are possibly not fragmented. Moreover, the masses finally measured are modified by various factors, such as the exact position of the fragmentation related to the peptidic bond, the number of charges on the ionic fragment, the possible loss of molecules (as water or ammonium) and the isotopic pattern of the peak. Another difficulty is the presence of possible modifications (adjunction of specific molecules on amino acids) or mutations in the source peptide, resulting in shifts in some of the measured masses.

Different algorithms aimed at correlating a theoretical peptidic sequence with an experimental MS/MS spectrum have been described in the literature [16, 15].

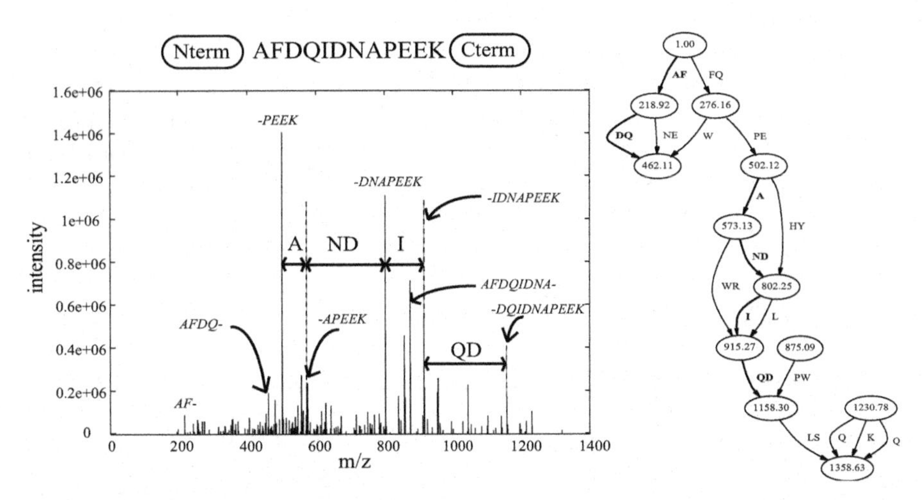

Fig. 2. Representation of an interpreted MS/MS spectrum and a simplified part of its associated spectrum graph. The corresponding source peptide sequence is AFDQIDNAPEEK. The tag ANDIQD is highlighted in the spectrum. Since this tag is formed by C-terminal fragments, it is reversed compared to the peptide sequence. Some algorithms, like Popitam, first transform all peaks into N-terminal fragments such that the tags can be read in the same direction. The first node of the spectrum graph corresponds to the empty sequence (mass of an hydrogen) and the last one corresponds to the complete sequence (parent mass of the MS/MS spectrum).

Our algorithm, called Popitam [18], exploits the concept of 'spectrum graph' in order to extract *tags* (amino acid sequences) from the MS/MS spectrum. The graph represents all possible complete sequences and sub-sequences that can possibly be built from the spectrum (see Fig. 2). Vertices are built from the peaks and represent therefore masses of fragments, while edges represent amino acid masses (consecutive fragmentation in the source peptide differ by the mass value of one amino acid). The sequences can be constructed by moving from one vertex to another by following existing edges. A more complete description of the spectrum graph has been published in [7]. The major difficulty with such a structure is the huge number of tags that can be extracted from the graph. Among them, only a tiny part represents true sub-sequences of the source peptide.

Identification methods based on tag search [7, 5, 26, 38, 40] typically try to extract tags (or complete sequences) from the graph and use them to identify the most similar sequence from the database using sequence alignment algorithms [25]. In Popitam, on the other hand, the database is used to direct the search in the graph. This has two major advantages: first, all the original information can be used during the comparison between the theoretical peptide and the graph, and second, the search space is strongly reduced, since tags are specifically extracted for each theoretical sequence.

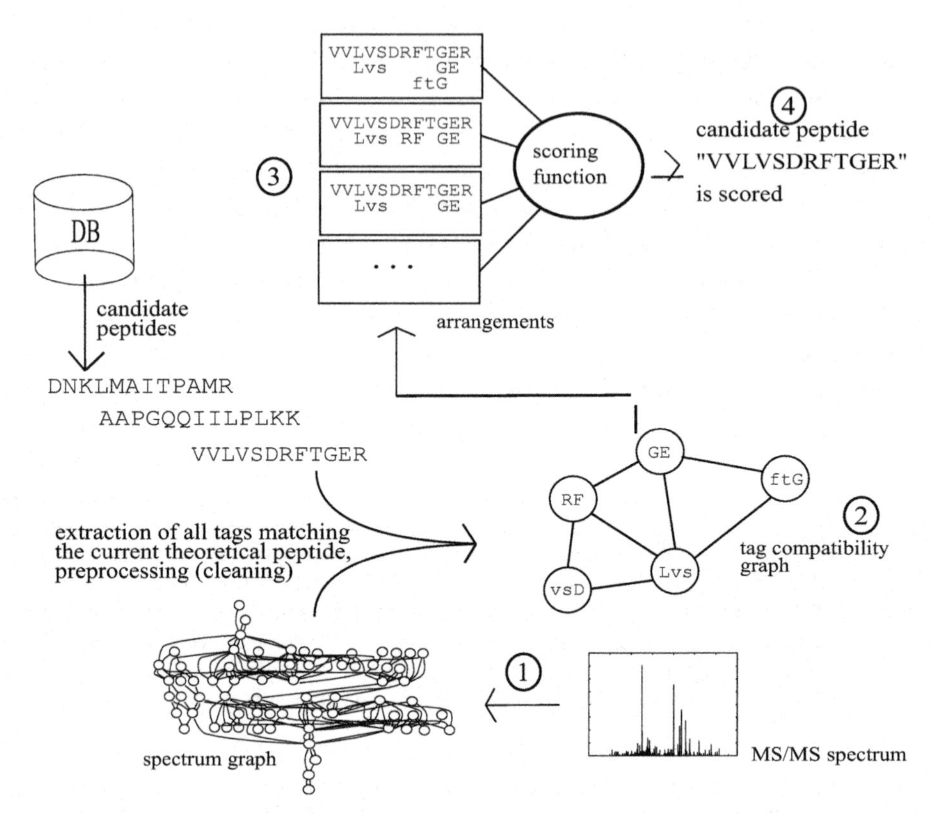

Fig. 3. Illustration of Popitam's algorithm. The first step consists in structuring the MS/MS spectrum into a spectrum graph (1). Then, theoretical peptide sequences guide the extraction of tags from the graph. As an example, five tags are extracted for the theoretical peptide VVLVSDRFTGER. The 'compatibility graph' (2) is build from the tag list and is used to compute all possible arrangements of tags (3), which are then scored. The best arrangement score is the final score of the theoretical peptide (4). All candidate peptides iteratively undergo the complete procedure from tag extraction to arrangement scoring. Eventually, the best-scored peptide is proposed as the most likely source sequence of the experimental MS/MS spectrum.

Popitam's algorithm can be described as follows (see Fig. 3: first, the experimental MS/MS spectrum is transformed into a spectrum graph. The next step is applied to each theoretical peptide from a database that matches some criteria, such as species of the sample and the molecular weight of the source peptide. It consists of extracting from the spectrum graph all tags with a sequence that matches a sub-sequence of the current theoretical peptide. The tags are then processed, in order to remove redundant information (i.e., tags that are sub-sequences of other tags are removed), and non-logical information (i.e., a tag which begins at the first node of the graph must also be prefix of the theoretical sequence). The aim is to arrange the tags in order to maximally score the theoretical peptide. In addition, arrangements have to take into account logical rules between tags. The arrangements are found by building a compatibility graph using these two rules, and then looking in the graph for all possible *cliques* (fully connected sub-graphs). The final step is to score each arrangement of the current theoretical peptide. The higher arrangement score represents the final identification score for the current peptide. This is of course a key process of Popitam's algorithm, since the arrangement scoring function should be maximal for the theoretical peptide that represents the source peptide (in order to automatically identify, our scoring function must ranks first the theoretical peptide that correspond to the source peptide), and should allow a good discrimination between the correct candidate and the other theoretical peptides. We defined for the scoring function nine sub-scores, such as the number of tags included in the arrangement or the percentage of coverage of the theoretical peptide by the tag. Other scores are described in [18]. Of note, new sub-scores may be added in the future, such as expert rules set by biologists used to studying MS/MS data.

We first combined empirically the nine sub-scores and tested this 'empirical' scoring function on a set of MS/MS spectra with known identifications. Then we used *Genetic Programming* (GP) [21] to discover a more efficient scoring function. GP belongs to the class of *evolutionary algorithms* (EA), which are stochastic search methods inspired by natural mechanisms. GP applies EA to a population of computer programs. A program can be represented as a tree with ordered branches in which the internal nodes are functions and the leaves are the so-called terminals of the problem. GP provides a way to search the space of all possible programs composed of functions and terminals in order to find a appropriate solution for the given problem. The evolutionary process starts with a population composed of random programs. Then this population applies the Darwinian principle of survival of the fittest and genetic mechanisms borrowed from biology to breed a new population of programs. This breeding process is repeated during a given number of generations in order to produce better and better approximations to an optimal solution of the given problem by exchanging 'genetic information' (BB) of promising points of the search space. The evolution is guided by a fitness function that determines how each program in the population solves the problem. In our case, trees represent the scoring function of the identification. Nodes are arithmetic and conditional operators, and leaves consist of the nine sub-scores computed for each tag arrangement and of constants randomly generated in a bounded set of values. For our problem, the fitness function indicates how a scoring function encoded by a program is able to correctly identify MS/MS spectra from a learning set. Genetic operators are the engines of the evolution. They allow producing offspring by combining and modifying the 'genetic information' contained in the individuals of the population. The most widely used operators are the crossover, the mutation and the permutation. They operate on parent trees that are selected from the population with, for instance, a *tournament selection* method [3]. The crossover operates on two parent trees.

The process begins by independently selecting a crossover point in each parent tree. Then the sub-trees, whose roots are a crossover point, are exchanged, giving rise to two offspring trees. The mutation operator consists of replacing a sub-tree of a parent tree by a new randomly generated sub-tree. Finally the permutation operator works by selecting a random internal node of a parent tree and permuting the order of its arguments.

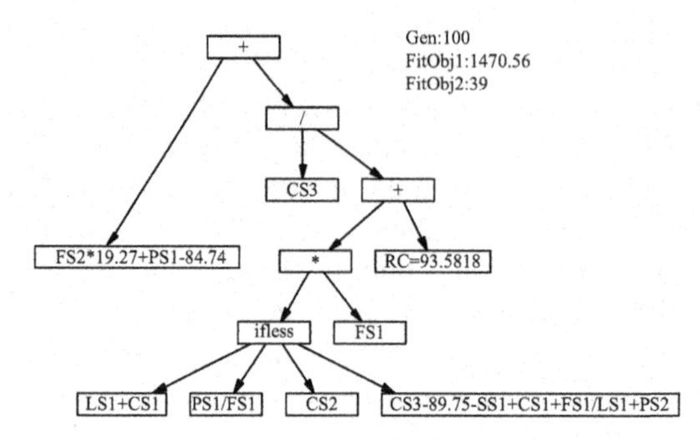

Fig. 4. An example of optimized scoring function learned by genetic programming, using a set of 265 MS\MS spectra. This function was then tested against a second set composed of 505 MS\MS spectra, and obtained a performance of 90.7% of correctly identified spectra, whereas the empirical function had a performance of 85.9%.

Using relevant setting of the genetic operators, of the fitness function, of the tree population size and of the generation number allows discovering new scoring functions with better performance for our problem (see Fig. 4).

3 A hierarchical cooperative multiple sequence alignment combining local similarity

The second major problem of proteomics that we handle concerns the analysis of proteic sequences in order to infer knowledge about the function of the corresponding proteins. In this section, we describe a cooperative multi-agent strategy that takes advantage of concerted cooperation to achieve a fully automated clustering of biological sequences depending on multiple local or distant similarities. Therefore, clustering leads to construction of a multiple sequence alignment (MSA) based on several independent alignments by regular single motifs (local conserved regions in several sequences, see Section 3.1) and linked dyad motifs (couple of covariated regions in several sequences, see Section 3.2).

During evolution, DNA sequences are subject to mutation. These mutations may have very different outcomes depending on where they occur. A point mutation that does not affect the survival of an organism is likely to remain and be passed on future generations. On the other hand, a mutation on a regulatory site may have dramatic consequences on the survival of the organism. Thus, this mutation is not likely to propagate to future generations. Considering a set of biological sequences known to be related (a set of promoters or a family of protein sequences for instance), common features or similar words (sub-sequences) shared by all the sequences can be determined. We suspect that these similar words have been kept by an evolutionary pressure, because they are involved in a biological process. Then the problem of discovering conserved regions can be formulated as follows: given a set of sequences, extract one word (of a constrained length) per sequence such that all extracted words share a maximum global similarity. These extracted words will then constitute a local multiple alignments. This problem is addressed in Section 3.1

The three-dimensional shape of a protein bears on the protein function. This shape is constrained by physico-chemical interactions, which determine the secondary and tertiary structures. Physico-chemical interactions are known, but it is very difficult to precisely predict their involvement in the folding of a linear sequence of amino acids into a structured protein. Motifs used in biology to characterize families of proteins are often single words that represent conserved regions in the corresponding set of related sequences. However, this kind of motif may not be descriptive enough because protein properties depend greatly on physico-chemical interactions between distant regions. Thus, a descriptive motif should at least be made up of regions linked by some dependence. This problem is addressed in Section 3.2.

MSA is a very difficult problem ever present in bioinformatics. The alignment of a collection of biological sequences significantly contributes to the field of protein characterization: it applies to phylogenetic analysis, structural modeling or functional annotation transfer from characterized to new sequences [29]. Several heuristic MSA algorithms have been developed so far, ranging from traditional progressive methods [8, 20, 41] to computationally expensive score optimization strategies [14, 28, 31, 39]. The quality of alignments produced is highly dependent on the initial data [22]. None of these methods performs well in the variety of situations encountered with biological data, for example low sequence identity, complex or inverted sequence similarity architecture, or variable sequence length.

3.1 A strategy for ungapped local multiple alignment

In this section, we address the following well-known problem: given a set of DNA or protein sequences known to be functionally related, choose exactly one fixed-length word per sequence, so that all chosen words are maximally similar (the length must be provided by the user). The set of chosen words is called an *ungapped local multiple alignment* (ULMA). This problem is addressed by a large variety of programs, but we only focus on those that use a similarity measure based on the information content concept. Considering that this problem has been shown to be NP-hard [1], all these methods are heuristics. Apart from CONSENSUS [19], which builds a solution using a greedy algorithm, other methods sample the search space, optimizing an objective function. The well-known programs are the Gibbs Site Sampler [23] and MEME [2]. Our method, MoDEL (motif discovering with evolutionary learning) [17], is based on the exploration of two search spaces using a linked optimization strategy: 1) the search space $\mathcal{M}$ consists of all possible words of a given length, and 2) the search space $\mathcal{P}$ consists of all possible ULMAs of this given length.

An evolutionary algorithm globally samples the word search space, producing sub-optimal ULMAs. These ULMAs are refined by two *hill-climbing operators*. The key idea is to use the evolutionary algorithm to detect relevant areas of the search space. We demonstrate some advantages of our strategy on a difficult proteins sequence data set (with a low level of local similarity). A web interface is available at `http://idefix.univ-rennes1.fr:8080/PatternDiscovery/`.

Definitions

Let $S = \{S_1; S_2; \ldots; S_N\}$ a set made up of N sequences of length T (for convenience, we suppose that all sequences have the same length). $S_{i,j}$ is the jth symbol of the ith sequence. S is over a fixed size alphabet $\Sigma = \{\sigma_1; \sigma_2; \ldots; \sigma_K\}$ with $K = |\Sigma|$ (4 for DNA and 20 for proteins). Let $M = \{m_1; m_2; \ldots; m_W\}$ be a word of length W defined on Σ. The set of all possible words M defines the search space $\mathcal{M}$ of size K^W. Let $P = \{p_1; p_2; \ldots; p_N\}$ be a position vector, defining occurrences on S_i, beginning at positions p_i and of length W. Thus, P represents an ungapped local multiple alignment (ULMA), made up of exactly one occurrence per sequence in S. The set of all possible ULMAs P defines the search space $\mathcal{P}$, of size $(T - W + 1)^N$.
The purpose of our method is to find the point of $\mathcal{P}$ that maximizes the *relative entropy*. This measure expresses how unexpected the symbol frequencies inside the ULMA columns are, with respect to some background frequencies (usually estimated from the whole dataset). In other words, more the symbols are conserved in the ULMA columns, and higher is this value. The relative entropy is calculated as follows: a frequency matrix F is first estimated from the ULMA: for $k = 1, \ldots, K$ and $j = 0, \ldots, W - 1$:

$$F_{k,j} = \frac{1}{N} \sum_{i=1}^{N} J\left(k, S_{i,p_i+j}\right) \quad \text{with} \quad J(k,a) = \begin{cases} 1 & a = \sigma_k \\ 0 & \text{else} \end{cases} \tag{1}$$

The relative entropy is calculated from this matrix by:

$$I\left(F, F^0\right) = \sum_{i=1}^{K} \sum_{j=1}^{W} F_{i,j} \log\left(\frac{F_{i,j}}{F_i^0}\right) \tag{2}$$

where F^0 is the background frequencies estimated once, during the initialization step, from S.
The exploration process takes advantage of both word and ULMA representations, mapping them together by two projection operators. The first one projects points from $\mathcal{M}$ onto $\mathcal{P}$ (MtoP) and the second one projects points from $\mathcal{P}$ onto $\mathcal{M}$ (PtoM). Both projection operators are illustrated in Fig. 5. The MtoP operator builds a ULMA from a word. This is achieved by aligning the word M with all the sequences of S. More formally: for $i = 1, \ldots, N$ and $j = 1, \ldots, T - W + 1$:

$$p_i = \underset{j}{\operatorname{argmax}} \sum_{k=1}^{W} H\left(S_{i,j+k-1}, m_k\right) \quad \text{with} \quad H(a,b) = \begin{cases} 1 & a = b \\ 0 & \text{else} \end{cases} \tag{3}$$

Then, p_i corresponds to the position on S_i where the match count is maximal. However, a single sequence may have several maximal positions, leading to a position vector having ambiguous p_i. In this case, we use a greedy algorithm to clear up ambiguities. Non-ambiguous dimensions are first selected to build a core. Remaining dimensions are then added one by one, choosing each time the position that maximizes the relative entropy of the growing ULMA. This operator has an interesting property: because of the alignments, the resulting occurrences in S present a much greater similarity than positions selected by chance (a random point of $\mathcal{P}$). This feature allows us to directly find outstanding points (with a high relative entropy value). To denote these outstanding points, we define $\mathcal{Q}$, a subspace of $\mathcal{P}$ which is made up by all points in $\mathcal{P}$ that can be reached from $\mathcal{M}$ by a projection. By construction, this subspace has at most the size of $\mathcal{M}$. The complementary operator, PtoM, produces a word from a ULMA. The resulting word is simply the most likely one according to the frequency matrix F (Equation 1). More formally, for $i = 1, \ldots, K$ and $j = 1, \ldots, W$:

$$m_j = \sigma_k \quad \text{with} \quad k = \underset{i}{\operatorname{argmax}} \left(F_{i,j} \right) \tag{4}$$

Note that the correspondence established between the two search spaces is not symmetrical. A word projected to $\mathcal{P}$ and projected back to $\mathcal{M}$ will not necessarily be the same as the initial one. This is also true for a ULMA.

Exploration

The overall strategy of MoDEL combines an evolutionary algorithm with simple hill-climbing optimizations. The evolutionary algorithm locally samples the $\mathcal{M}$ search space. Genetic operators are thus applied on words (points of $\mathcal{M}$). In addition to classical genetic operators (crossover, mutation), specific ones have been designed. We use for example a slide operator, which shifts all symbols in the word one position to the left or to the right. The fitness value of a given $\mathcal{M}$ point is the relative entropy of the corresponding $\mathcal{Q}$ point (which is obtained by the MtoP projection operator). This evolutionary algorithm allows to discover high-scoring ULMAs belonging to the $\mathcal{Q}$ search space. However, exploring solely this space is not sufficient because the global maximum might be located in an area of $\mathcal{P}$ that is not covered by $\mathcal{Q}$. A local search in $\mathcal{P}$ is necessary to locate possible higher scoring ULMAs. After each iteration of the evolutionary algorithm, about 5% of the newly created solutions are chosen to be optimized with two simple hill-climbing processes: a 'sequence-neighborhood' optimization and a phase-shift correction. The sequence-neighborhood works as follows: given a position vector $P = \{p_1, p_2, \ldots, p_N\}$, only one p_i dimension is modified at once, keeping the others fixed. This dimension is sampled on all possible values (corresponding to the $T - W + 1$ possible words of S_i), and is updated with the position that maximizes the overall relative entropy (equation 2). All dimensions are modified one after the other, either in a predefined order or by random selection without replacement. The sequence-neighborhood optimization is coupled with a phase shift

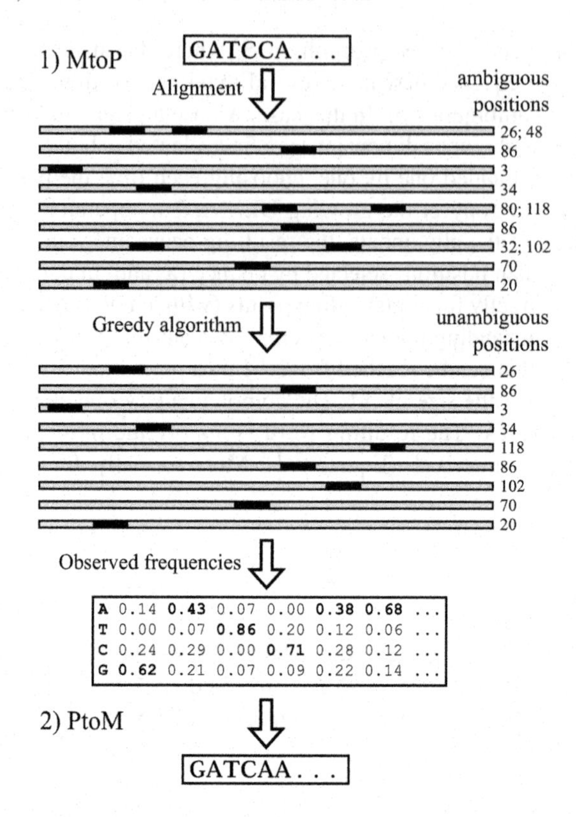

Fig. 5. Projection operators allow the mapping of a point from one search space to the other. 1) MtoP: the word (a point of $\mathcal{M}$) is aligned on each sequence in order to find positions that maximize the match count. It is likely that several positions on the same sequence will maximize the match count, leading to a position vector having some ambiguous positions. A greedy algorithm clears these ambiguities up. Non-ambiguous dimensions are first selected to build a core. Remaining dimensions are then added one by one, each time choosing the one that maximizes the relative entropy of the growing ULMA. 2) PtoM: the resulting word is the most likely one according to the frequency matrix F.

correction [23], which consists of shifting all p_i together, several positions to the left and to the right to ensure that the ULMA is not in a shifted sub-optimal configuration.

Performance comparison of exploration strategies

One may ask if using a complicated exploration strategy really gives an advantage over simple repeated optimizations (multi-start hill-climbing) on different random seeds, when the same CPU-time is allowed. To illustrate the relevance of the global search (evolutionary algorithm) as well as the MtoP projection operator, we compared the exploration abilities of MoDEL versus three simpler strategies. The first one consists simply of local hill-climbing optimizations of random points of $\mathcal{P}$. The second one works in the same way but with random points of $\mathcal{Q}$ (by converting random words to ULMAs using the MtoP operator). The last one is the Site Sampler, which can be roughly seen as a stochastic hill-climbing from random points of $\mathcal{P}$. We used a data set of 16 helix-turn-helix (HTH) protein sequences. This structural motif (HTH) consists of two α-helices separated by a short turn. Its conformation allows proteins to specifically bind DNA. The second helix recognizes and binds to a specific sequence along the major groove of DNA, while the first helix stabilizes the complex. We build a data set by gathering 16 HTH domain-containing proteins, each one being a representant of a major HTH evolutionary lineage [37]. In this way, sequences are highly divergent and the HTH occurrences are weakly similar, and quite difficult to recover. The Swiss-Prot database [4] accession numbers of these sequences are: P03033, P00579, P22874, P02306, P03809, P16528, P16644, P03020, P46833, P43271, P03030, P30338, P03021, P06020, P09964, P23221. The highest-scoring ULMA we found (corresponding to the HTH domains) has a relative entropy of 32.71 (we fixed a width of 20 amino acids). We will denote this ULMA as the reference. Fig. 6 shows the success rate (y-axis) plotted against the CPU-time (x-axis). The success rate is estimated over 100 independent runs. A success means that the reference ULMA has been recovered. Results show that the evolutionary algorithm significantly improves the exploration process, compared to simple hill-climbing optimizations from random seeds. We therefore show that, even if the evolutionary algorithm spends about 50% of the CPU-time of MoDEL, it gives a significant advantage in the exploration process, by choosing relevant points of $\mathcal{Q}$, to be locally optimized.

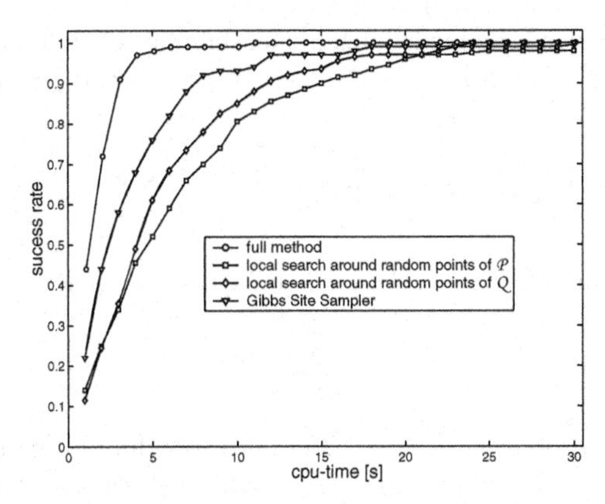

Fig. 6. Performance of MoDEL and simpler strategies on the HTH dataset. The success rate is plotted against the CPU-time (Athlon, 1.6 GHz). 1) Local search around random points of $\mathcal{P}$: ULMAs are randomly sampled and optimized with the hill-climbing optimizations. 2) Local search around random points of $\mathcal{Q}$: same as above but the MtoP projection is used before the hill-climbing optimizations. 3) Full method: the $\mathcal{Q}$ search space is sampled with the evolutionary algorithm, which chooses points to be locally optimized. 4) Gibbs Site Sampler.

3.2 Linked dyad motif discovery from combination of conserved regions

The biological mechanisms underlying the three-dimensional folding of proteins are not sufficiently known yet to be used efficiently for the discovery of structural motifs (i.e., motifs taking physico-chemical interactions into account). We focused on the idea of using covariation measurement to detect a statistical dependence between conserved regions in a set of sequences. Thus, we hope that regions correlated in this way have a real biological meaning [10, 36]. We define a *linked dyad* as a pair of conserved ULMA with a significant covariation value between them. In linguistic terms, linked dyads belong to the upper class of context dependent languages in Chomsky's hierarchy and thus are considered very difficult (NP-hard) to discover. Our goal is to discover such motifs from a given sequence set, and because it is unrealistic to achieve that by an exact method, we will use an evolutionary cooperative metaheuristic based on concepts defined in MoDEL.

We have a set of sequences using alphabet Σ and two position vectors (see Section 3.1) of length N (with N the number of considered sequences), we define $M(P1, P2)$ as the well known *mutual information* measure between the two vectors $P1 = \{P1[1], \cdots, P1[N]\}$ and $P2 = \{P2[1], \cdots, P2[N]\}$, slightly modified for our problem.

$$M\left(P_1, P_2\right) = \sum_{x,y} \left(f_{xy}\left(P_1, P_2\right) \cdot \log \frac{f_{xy}\left(P_1, P_2\right)}{f_x\left(P_1\right) \cdot f_y\left(P_2\right)} \right)$$

Where $f_x\left(P_1\right)$ is the frequency of letter x in P_1 and $f_{xy}\left(P_1, P_2\right)$ is the frequency of co-occurrence of x and y in the same sequence.

This definition can be extended that to measure the covariation between two conserved patterns $P_1^{1\cdots W}$ and $P_2^{1\cdots W}$ of same size W beginning at position vectors P_1 and P_2:

$$C_L(P_1, P_2) = C(P_1^{1\cdots W}, P_2^{1\cdots W}) = \sum_{i=1\cdots W} M(P_1^i, P_2^i) \tag{5}$$

Where $P_1^1 = P_1$ and $P_1^i[k] = P_1[k] + i - 1$ We also compute

$$\overline{C_L}(P_1, P_2) = \sum_{i=1\cdots W} M(P_1^i, P_2^{L-i+1}) \tag{6}$$

which takes into account the reverse order corresponding to 'palindromic' covariation. We then select pairs of position vectors with high value of C_L or $\overline{C_L}$.

Evolutionary strategy for linked dyad discovery

MoDEL is efficient to find the best conserved region of a set of sequences, but the diversity of the population obtained after several generations is poor: weaker chromosomes are discarded during the evolutionary process and the resulting solutions are always slight variations of the best one. To avoid that, we use the concept of ecological niches [11]. Each chromosome fitness is balanced with a distance between the chromosome and the rest of the population. Thus the fitness of a 'weak' chromosome with a few neighbors (similar chromosomes) in the population is as good as that of a 'strong' chromosome with many neighbors. So the population will be distributed among many optima, and not only around one unique optimum like in MoDEL.

With this modification the MoDEL evolutionary process finally yields several different conserved regions. If we use MoDEL as a generator of conserved regions, we can compute the mutual information C_L and $\overline{C_L}$ between all possible pairs of such regions and keep the best pairs. Depending on the number of possible pairs, conserved regions can be combined via a heuristic step.

Preliminary results

In order to validate our hypothesis on mutual information measurement, we have implemented a generator which produces sequences similar to those of the Pevzner's challenge [34]: find a signal in a sample of sequences, each 600 nucleotides long and each containing an unknown signal (pattern) of length 15 with 4 mismatches. In our generator, instead of regular motifs, linked dyads are inserted. Given two consensus words, one being a morphic transformation of the other one, they are inserted in each sequence with a given number of errors. Then, for each occurrence of these words, we randomly restore the morphic links, which were lost because of errors. Any mutation in one word will be reflected onto the other one, creating two sets of words complying with our definition of a linked dyad. We have used this generator to create sets of 20 sequences of length 1 000, with 10 inserted linked dyads of length 15 (for each word) having variable rates of errors and mutual information values.

We used MoDEL (modified as described above) on these sets of sequences, and calculated the mutual information values of all possible pairs composed of regions generated by the evolutionary process. Note that we gave the chromosomes the appropriate length; with a smaller (resp. greater) length, we would have retrieved regions that are included (resp. contained) in the inserted regions. The covariation computation is very time consuming if we consider all the pairs, but is reduced to a few seconds with a heuristic approach selecting the most promising pairs.

3.3 A cooperative multiple sequence alignment based on biological domain composition

MSA requires a very flexible heuristic to produce a biologically meaningful alignment, keeping time and computation resources within reasonable ranges and preserving robustness with regard to the initial data set (the number of sequences, the degree of similarity between sequences, the variability in domain composition and organization). Several difficulties are encountered: the choice of a pertinent scoring system and the efficient exploration of the search space, which is exponential to the number of sequences. MSA is therefore highly combinatorial and requires strong heuristic strategies since exhaustive methods are impracticable, even for reasonably sized data (three sequences or less) [30].

A concerted cooperative multi-agent approach, which is distributed and highly adjustable, is appropriate in this situation. The particularity of our strategy rests on the application of a preliminary clustering stage based on biological domains before constructing the alignment itself. The heuristics lies in splitting the problem into a variable number of agents that consider only a single domain at a time and achieve a dynamic assembly of sequences around this domain. Concerted cooperation is driving the clustering process. This strategy, unlike traditional clustering methods, does not require to pre-set the number of clusters and tolerates overlapping clusters, it is therefore robust with regard to starting conditions.

Since this approach is entirely founded on the notion of domains, it should be able to take advantage of any single piece of information contained in biological domains. The term 'biological domain' covers here not only regular motifs (Section 3.2), but also linked dyad motifs (Section 3.3). Moreover, domains found only in a sub-set of sequences are also considered. The order in which the different domains are organized in the sequences do not impair the procedure. Basic concepts, and a first draft of this novel MSA algorithm as well as preliminary results are presented below.

Clusters and sequences scoring system relies on domain information

The dependency on pairwise similarity, either global or local, is a major weakness of the existing sequence clustering or alignment methods. The global measure is not optimal to describe the homogeneity of a group of biological sequences, since it does not reflect their domain architecture, which is fundamental to their biological function. We propose a similarity criterion that does not rely on pairwise comparisons but includes the relative entropy of a regular motif inferred across all sequences on the one hand (Section 3.1, equation 2); and the co-variation measure of linked dyad motifs on the other hand (Section 3.2, equations 5 and 6). This two-component similarity criterion is local and hence it accounts for complex domain architecture with potential inversions. In addition, it embodies simultaneously the conservation of all considered sequences. Therefore, it is more suitable than a pairwise score to characterize homogeneity within a group of sequences. Finally, the contribution of each sequence to this similarity measure is computed and serves as a basis for sequence exchange between cooperating agents or clusters during the optimization process.

Cooperative agents explore of the search space

Exploration of the partition space, especially when overlapping clusters are authorized, is an NP-hard problem. To bypass this situation, we propose a cooperative heuristic that distributes the exploration among a variable number of agents, each agent being responsible for one cluster. The optimization of a single sequence family can be properly addressed by a single agent. Each agent cooperates with other agents to perform a local optimization on the corresponding sequence family. Concerted cooperation is a decision-based communication procedure that results in sequence exchange between agents. This approach can be seen as a balance between two procedures: intensification and diversification of the search. Intensification tends to focus on promising points of the search space, whereas diversification directs the search towards yet unexplored points.

Intensification implies small moves in the search space. It is mostly achieved by concerted cooperation, which consists of sequence exchange and is decided by two partners according to their size, fitness and sequence scores. Upon agreement between two partners, a certain exchange is executed. Diversification induces more radical moves in the search space. It encompasses birth of new agents, death of incompetent agents, fission or fusion of agents. Agents have also the possibility of picking randomly new sequences from the initial set, to maintain some diversity. Inversely, they can exclude bad sequences to avoid uncontrolled growth of the family and speed the convergence towards a stable sequence cluster.

Once the clustering process has reached equilibrium, a multiple alignment can be constructed. The sequence family of each agent embodies a building block of the whole cluster set. During the optimization process, agents self-organize into a hierarchical structure. The assembly of all families guided by the natural hierarchy present among agents will lead to an emerging multiple sequence alignment.

First draft

In the present chapter, we describe a first draft of the clustering algorithm involving cooperative agents. A parallel implementation has been performed using the 'message passing interface' package MPI [32]. Once the agents are initialized, they run a number of MoDEL generations to compute a preliminary best motif with its relative entropy, set as the agent's fitness. The contribution of each sequence to the agent's fitness is also computed. These scores and the best motif found so far are sent to all other agents so that they acknowledge each other. The agents check the information received from other agents, if any and update their data. Then a different communication procedure prompts the agent to decide which exchange should take place with which partner and actually sends the sequences. When this second communication is over, the agent looks for putative received sequences. It decides whether or not to pick new sequences or trash some poor scoring sequences. Finally, the agent updates its sequence set and starts again the succession of actions of the algorithm.

The decisions to exchange or to pick or trash sequences bring a certain benefit (or detriment) to the agent's fitness. This benefit is used to choose which type of decision should be privileged in future iterations.

Our clustering approach was tested on artificially-generated DNA (cardinal 4) sequence families in which a conserved motif of length 15 with 2 errors was inserted [34]. A set of 100 sequences, 5 families (5 different motifs) of 20 sequences of length 600, was produced. A population of 14 agents received each a random subset of 46 sequences from the initial set. The subset was iteratively updated during the clustering process, by sequence exchange between agents as described in Section 3.4.3. A total of 50 iterations requires 71 seconds CPU-time when distributed on a cluster of 14 Pentium4 1.5GHz processors. This whole clustering procedure was repeated 100 times on different data sets for statistical reasons. The results obtained were the following: among the 5 families contained in the initial set, a mean of 4.31 families was identified by the agents. At the end of the procedure, each cluster contained a mean of 52.2% of sequences from its identified family, the remaining was distributed among the 4 other families. Finally, each cluster contained on average 69.5% of the sequences from its identified family (14 sequences out of 20).

Although this version of the algorithm is still under development, it produced encouraging results, confirming the suitability of the scoring system.

4 Conclusion

We presented in this paper a new classification of metaheuristics considering the property of cooperation among entities exploring a search space. This classification involves three sub-classes: centralized, individual and concerted cooperation depending on how the cooperation is accomplished. We gave new definitions and described new developments of these approaches. We emphasized the importance of a combination of these methods to maximally profit from their specific characteristics for complex structured problems. In this case, a hierarchical division of the problem into independent tasks can lead to spread and simplify optimization steps. The results of these optimizations are then associated to produce a global solution benefiting from the structures appearing in the different levels considered. We applied all these techniques to two central problems of proteomics: automatic protein identification and multiple sequence alignment based on motif discovery. We gave a short overview of the state of the art of these problems and some possible improvements to manage all their inherent difficulties. We gave also some promising preliminary results obtained with our tools using real data for protein identification, learning of a discriminating score, motif inference or sequence clustering.

We are currently working on several enhancements of our methods. From the application point of view, we try to integrate some new biological expert knowledge in the parsing of the graph structure and in the protein identification scoring function. We also designed a new algorithm based on tag matching which will allow the detection of all possible modifications. We work on the extension of motif representation allowing variable length, gaps inside motif or multiple alphabet representation. A wide variety of improvements are also currently under study for the metaheuristic aspects. We work on a better management of the building block concept in evolutionary algorithms. For example, if we can represent explicitly the possible building blocks in genetic programming then we can use a multi-level parallel algorithm to discover and combine them. We will extend our explored space representation to control the use of the genetic operators and to define smarter new operators for genetic algorithms. We will develop adaptive strategies for agents that enable a mutual agreement and increase the diversification aspect of concerted cooperation using new operators like birth or death of agents. Finally, we plan to integrate a multi-objective optimization approach [6] to our methods because most of the biological problems imply several independent properties with inconsistent maximum values.

References

1. T. Akutsu, H. Arimura, and S. Shimozono. On approximation algorithms for local multiple alignment. Proceeding 4th Int. Conf. Computational Molecular Biology, pages 1–7, 2000.
2. T.L. Bailey and C. Elkan. Fitting a mixture model by expectation maximization to discover motifs in biopolymers. Proceedings of the Second International Conference on Intelligent Systems for Molecular Biology., pages 28–36. AAAI Press, 1994.
3. T. Blickle and L. Thiele. A comparison of selection schemes used in genetic algorithms. Technical Report TIK 11, 1995.
4. B. Boeckmann, A. Bairoch, R. Apweiler, M.-C. Blatter, and A. Estreicher et al. The SWISS-PROT protein knowledgebase and its supplement TrEMBL in 2003. *Nucleic Acids Res.*, 31:365–370, 2003.
5. T. Chen, M.Y. Kao, M. Tepel, J. Rush, and G.M. Church. A dynamic programming approach to de novo peptide sequencing via tandem mass spectrometry. *J. Comput. Biol.*, 8(3):325–337, 2001.
6. C.A. Coello Coello, D.A.V. Veldhuizen, and G.B. Lamont. *Evolutionary Algorithms for Solving Multi-Objective Problems.* Kluwer Academic Publisher, 2002.
7. V. Dancik, T. Addona, K. Clauser, J. Vath, and P.A. Pevzner. De novo peptide sequencing via tandem mass spectrometry. *J. Comput. Biol.*, 6:327–342, 1999.
8. D.F. Feng and R.F. Doolittle. Progressive sequence alignment as a prerequisite to correct phylogenetic trees. *J.Mol.Evol.*, 25(4):351–360, 1987.
9. F. Fernandez, M. Tomassini, W.F. Punch III, and J.M. Sanchez. Experimental study of multipopulation parallel genetic programming. Proceedings of the Third European Conference on Genetic Programming, pages 283–293. Springer Verlag, 2000.
10. K. Fukami-Kobayashi, D.R. Schreiber, and S.A. Benner. Detecting compensatory covariation signals in protein evolution using reconstructed ancestral sequences. *J. Mol. Biol.*, 319(3):729–743, 2002.
11. D.E. Goldberg. *Genetic Algorithm in Search, Optimization and Machine Learning*. 1989.
12. D.E. Goldberg. *The Design of Innovation: Lessons from and for Competent Genetic Algorithms*. Kluwer Academic Publishers, 2002.
13. W. Golubski. Genetic programming: A parallel approach. *Lecture notes in computer science*, 2311, 2002.
14. O. Gotoh. Significant improvement in accuracy of multiple protein sequence alignments by iterative refinement as assessed by reference to structural alignments. *J. Mol. Biol.*, 264(4):823–838, 1996.
15. R. Gras, P. Hernandez, M. Muller, and R.D.Appel. Scoring functions for mass spectrometric protein identification. In *Handbook of Proteomic Methods*, pages 477–485. Humana Press, 2003.
16. R. Gras and M. Muller. Computational aspects of protein identification by mass spectrometry. *Current Opinion in Molecular Therapeutics*, 3(6):526–532, 2001.
17. D. Hernandez, R. Gras, and R.D. Appel. MoDEL: An efficient strategy for ungapped local multiple alignment. *Computational Biology and Chemistry*, 28(2):119–128, 2004.
18. P. Hernandez, R. Gras, J. Frey, and R.D. Appel. Popitam: Towards new heuristic strategies to improve protein identification from tandem mass spectrometry data. *Proteomics*, 3(6):870–879, 2003.
19. G.Z. Hertz and G.D. Stormo. Identifying dna and protein pattern with statistically significant alignment of multiple sequence. *Bioinformatics*, 15(7/8):563–577, 1999.
20. D.G. Higgins and P.M. Sharp. Fast and sensitive multiple sequence alignments on a microcomputer. *Comput.Appl.Biosci.*, 5(2):151–153, 1989.

21. J.R. Koza. *Genetic Programming: on the programming of computers by means of natural selection.* The MIT Press, 1992.
22. T. Lassmann and E.L. Sonnhammer. Quality assessment of multiple alignment programs. *FEBS lett.*, 529(1):126–130, 2002.
23. C.E. Lawrence, S.F. Altschul, M.S. Boguski, J.S. Liu, and A.F. Neuwald et al. Detecting subtle sequence signals: a gibbs sampling strategy for multiple alignment. *Science*, 262(208):214, 1993.
24. S.C. Lin, W.F. Punch III, and D. Goodman. Coarse-grain parallel genetic algorithms: Categorization and new approach. volume Sixth IEEE parallel and distributed processing, pages 28–37, 1994.
25. A.J. Mackey, T.A. Haystead, and W.R. Pearson. Getting more from less: algorithms for rapid protein identification with multiple short peptide sequences. *Mol. Cell Proteomics*, 1:139–147, 2002.
26. M. Mann and M. Wilm. Error-tolerant identification of peptides in sequence databases by peptide sequence tags. *Anal Chem*, 66:4390–4399, 1994.
27. Z. Michalewicz and D. Fogel. *How to Solve It: Modern Heuristics.* Springer-Verlag, 2000.
28. B. Morgenstern. DIALIGN 2: improvement of the segment-to-segment approach to multiple sequence alignment. *Bioinformatics.*, 15(3):211–218, 1999.
29. L.J. Mullan. Multiple sequence alignment–the gateway to further analysis. *Brief.Bioinform.*, 3(3):303–305, 2002.
30. C. Notredame. Recent progress in multiple sequence alignment: a survey. *Pharmacogenomics*, 3:131–144, 2002.
31. C. Notredame and D.G. Higgins. Sequence alignment by genetic algorithm. *Nucleic Acids Res.*, 24(8):1515–1524, 1996.
32. P.S. Pacheco. *Parallel programming with MPI.* Morgan Kaufmann, 1997.
33. S.R. Pennington and M.J. Dunn. *Proteomics from protein sequence to function.* BIOS Scientific, 2001.
34. P.A. Pevzner and S.-H. Sze. Combinatorial approaches to finding subtle signals in dna sequences. Proceedings of the eighth International Conference on Intelligent Systems for Molecular Biology, pages 269–278, San Diego, 2000.
35. W.F. Punch. How effective are multiple populations in genetic programming. Genetic Programming, pages 308–313, 1998.
36. D.J. Rigden. Use of covariance analysis for the prediction of structural domain boundaries from multiple protein sequence alignments. *Protein eng.*, 15(2):65–77, 2002.
37. J.A. Rosinski and W.R. Atchley. Molecular evolution of helix-turn-helix proteins. *J. Mol. Evol.*, 49:301–309, 1999.
38. A. Schlosser and W.D. Lehmann. Patchwork peptide sequencing: extraction of sequence information from accurate mass data of peptide tandem mass spectra recorded at high resolution. *Proteomics*, 2:524–533, 2002.
39. J. Stoye. Multiple sequence alignment with the divide-and-conquer method. *Gene*, 211(2):GC45–GC56, 1998.
40. J.A. Taylor and R.S. Johnson. Sequence database searches via de novo peptide sequencing by tandem mass spectrometry. *Rapid Commun Mass Spectrom*, 11:1067–1075, 1997.
41. J.D. Thompson, D.G. Higgins, and T.J. Gibson. CLUSTAL W: improving the sensitivity of progressive multiple sequence alignment through sequence weighting, position-specific gap penalties and weight matrix choice. *Nucleic Acids Res.*, 22(22):4673–4680, 1994.
42. M.R. Wilkins, K.L. Williams, R.D. Appel, and D.F. Hochstrasser. *Proteome Research: New Frontiers in Functional Genomics.* Springer-Verlag, 1997.

Integrating Gene Expression Data, Protein Interaction Data, and Ontology-Based Literature Searches

Panos Dafas[1], Alexander Kozlenkov[1], Alan Robinson[2] and Michael Schroeder[3]

[1] Department of Computing, City University, London, UK.
E-mail: {panos,a.kozlenkov}@soi.city.ac.uk
[2] MRC Dunn Human Nutrition Unit, Cambridge, UK.
E-mail: ajr@mrc-dunn.cam.ac.uk
[3] Biotec/Dept. of Computing, TU Dresden, Germany.
E-mail: ms@mpi-cbg.de

Summary. Until recently, genomics and proteomics have commonly been separate fields that are studied and applied independently. We introduce the BioGrid[4] platform, which aims to bridge this gap by integrating gene expression and protein interaction data. In the expression space, gene expression data can be analyzed using standard clustering techniques. To link the gene expression space with the protein interaction space, we assign domains and superfamilies to gene products by applying the *SUPERFAMILY* tool and the *Structural Classification of Proteins* (SCOP) database. For these proteins, the BioGrid platform may display possible physical interactions between them as predicted by the *Protein Structure Interactome Map* (PSIMAP). Any findings in the gene expression and protein interaction space should be compared with those reported in the scientific literature. Therefore both spaces are linked to a literature space by integrating GoPubMed, a system for ontology-based literature searches, into the BioGrid platform. We illustrate the approach that the BioGrid platform enables through an analysis of energy-related genes and protein complexes.

1 Introduction

Bioinformatics acquired genomics as one of its core fields of application after many complete bacterial genomes were sequenced around the mid 1990s. For the complete understanding of individual proteins and their functions, the technologies of proteomics are critically important. The experimental measurement of gene and protein expression levels has produced preliminary results on the regulation, pathways and networks of genes in cells. The ultimate aim of both genomics and proteomics in a bioinformatics and systems biology perspective is to map out all the circuits of energy and information processing in life. There are two initial challenges in systems biology and bioinformatics: one is to produce precise and accurate experimental data using mass spectrometer, protein chips and microarrays on the expression and

[4] EU project BioGrid (IST-2002-38344), http://www.bio-grid.net/.

W. Dubitzky and F. Azuaje (eds.), Artificial Intelligence Methods and Tools for Systems Biology, 107–127.

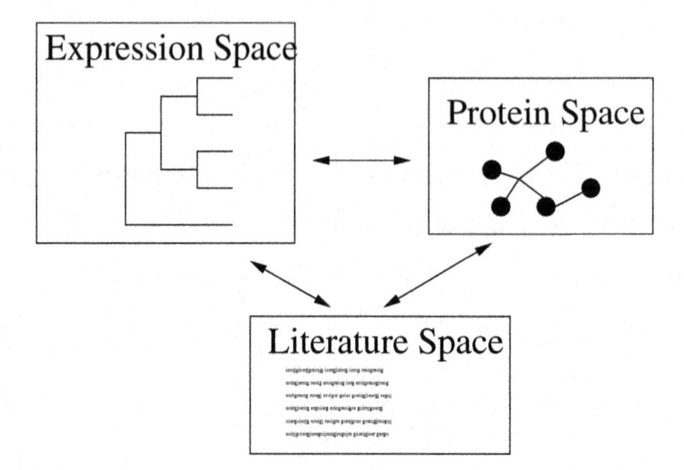

Fig. 1. The BioGrid platform integrates an expression space to analyse gene expression profiles, an interaction space to study protein interactions, and a literature space to find and classify relevant scientific literature.

interaction of genes and proteins in cells. The other is to organize these data in the most insightful and biologically relevant way so that the most information may be extracted and validated. We are addressing the second challenge by developing the BioGrid platform, which will allow biomedical researchers to easily understand, navigate, and interactively explore relationships and dependencies of genes and proteins by integrating data and analysis methods for gene expression, protein interaction and literature data (see Fig. 1). The platform enables users to cluster gene expression profiles, and to predict the domains, superfamilies and interactions of proteins. The data analysis is complemented by a novel ontology-based literature search tool, GoPubMed, that classifies collections of papers from literature searches into a navigable ontology.

2 Expression space

2.1 Gene expression data

The advent of DNA chip technology [11] facilitates the systematic and simultaneous measurement of the expression levels of thousands of genes. Consider Table 1, which shows data from experiments to identify all the genes whose mRNA levels are regulated by the cell cycle of yeast —the tightly controlled process during which a yeast cell grows and then divides [42, 12]. Each gene is characterized by a series of expression measurements taken at successive time intervals. The alpha cell experiment contains 21 successive measurements on samples taken from the same cell population. From the complete set of genes, the authors selected 800 genes whose level of expression fluctuates with the period of the cell cycle. We are thus dealing with a multivariate analysis of a data matrix with 800 rows (entries) and 21 columns (variables).

Table 1. Fragment of a multivariate data table of gene expression measurements. Rows correspond to genes, and columns to different experiments. The expression level of *circa* 6 000 genes was measured using microarray analysis at 21 successive time points by taking samples every seven minutes from a population of synchronized yeast cells [12].

Gene	0 min.	7 min.	14 min.	21 min.	28 min.	35 min.	...
YER150W	0.41	1.47	1.8	0.81	0.03	-0.31	...
YGR146C	0.78	0.37	-0.09	0.07	0.03	0.25	...
YDR461W	2.36	2.35	2.3	2.11	1.75	0.76	...
...							

A set of expression measurements for a gene is commonly referred to as the expression profile of that gene. To understand their gene expression data, scientists often wish to analyze and visualize it using a tool that groups genes with similar expression profiles in order to detect clusters of genes which are probably involved in a common biological process. Clustering can be defined as the process of automatically finding groups (i.e., clusters) of similar items based on some characteristics describing those items. Clustering is commonly used to infer information about the function of uncharacterized genes by applying the 'guilt by association' principle, i.e., if an uncharacterized gene is clustered with a group of genes participating in a known biological process (e.g., cell death or protein degradation), then it is assumed that the uncharacterized gene also participates in this process. However this information is often quite noisy because a given expression profile does not imply a given molecular function or biological process, however the reverse case is usually true. In many cases, gene expression profiles can be represented as a time series, so the problem of clustering gene expression profiles corresponds to identifying and grouping similar time-series data into clusters.

It is worth noting that the aim of clustering expression data is the same as the general aim of all data mining in bioinformatics and systems biology: finding the effects that unknown and hidden dynamics have on the expression profiles. However the biological interpretation of such results is an extremely hard procedure and should be always backed up by strong and robust biological arguments, and confirmed by laboratory experiments.

Clustering of gene expression data may be broken down into two steps:

1. Define a distance metric that measures the similarity of gene expression profiles.
2. Use these distances to group similar expression profiles together.

2.2 A catalogue of distance and dissimilarity/similarity measures

Starting from a raw data set of gene expression results, the first and most important step is to define a distance or (dis)similarity measure between the different gene expression profiles. This measure will determine which genes will be considered related and hence clustered together, and thus influence the subsequent analysis significantly. The following distance measures are useful (as a reference, see e.g., [29]):

Before we define the distances, let us define the scalar product and the Euclidean norm. Let $x, y \in \mathcal{R}^n$ be vectors. Then $(x, y) = \sum_{i=1}^{n} x_i y_i$ is called scalar product and $||x|| = \sqrt{\sum_{i=1}^{n} x_i^2}$ Euclidean norm. The mean value of x is given by $\mu_x = \frac{1}{n} \sum_{i=1}^{n} x_i$.

1. Scalar product: $d_c(x, y) = \sum_{i=1}^{n} x_i y_i$.
2. Maximum scalar product: $d_{c_{max}}(x, y) = max_k \sum_{i=1}^{n} x_i y_{i-k}, \quad -n \leq k \leq n$.
3. Direction cosine: $d_{cos}(x, y) = cos\theta = \frac{(x,y)}{||x||||y||}$ and angle (angular distance): $d_{angle}(x, y) = acos(d_{cos}(x, y))$
4. Correlation metric: $d_{cor}(x, y) = 1 - \frac{\left(\sum_{i=1}^{n}(x_i - \mu_x)(y_i - \mu_y)\right)^2}{\sum_{i=1}^{n}(x_i - \mu_x)^2 \sum_{i=1}^{n}(y_i - \mu_y)^2}$
5. Euclidean distance: $d_e(x, y) = ||x - y|| = \sqrt{\sum_{i=1}^{n}(x_i - y_i)^2}$
6. Minkowski distance: $d_m(x, y) = (\sum_{i=1}^{n} |x_i - y_i|^{\lambda})^{\frac{1}{\lambda}}, \lambda \in \mathcal{R}$

The above distance measures are not equivalent and the validity of any interpretation depends crucially on the choice of an appropriate metric. For example, Euclidean distance should be avoided, because it depends on the absolute level of expression and it is common to find genes which are co-regulated, in the sense that they respond to the environmental conditions in the same way, but with very different absolute levels of expression. The correlation coefficient is a better estimator of co-regulation than the Euclidean distance. However, this metric suffers from the opposite weakness: it is totally independent of the absolute levels. Consequently, strong correlations might be established between genes which are not co-regulated, but show small random fluctuations in expression that by chance exhibit a statistically significant correlation. The dot product or preferably the co-variance, seems the most appropriate to measure the co-regulation of two genes without over-estimating the weakly regulated genes [40].

2.3 Clustering of the data

Now let us consider two widely used clustering methods. For a more comprehensive overview on clustering, classification and visualization of gene expression data, see [40, 55, 21, 34, 14, 54, 30, 50].

K-means clustering

K-means clustering is one of the simplest and most popular clustering techniques. The algorithm is given the desired number of clusters, and empty clusters are formed, whose centroids are either distributed evenly across the domain space, or are randomly chosen elements from the set of points. Points are then assigned to the closest cluster, and centroids are recalculated. Clusters are emptied and the process is repeated until the assignments become constant. The main advantage of K-means lies in its simplicity and intuitiveness, as well as in its speed. When dealing with gene expression data, points to be clustered are n-dimensional, where each of the n values represent the expression level of the specific gene under given experimental conditions. One downside in this instance is that the number of clusters must be specified in advance, something that may require a certain amount of experimentation before it produces optimum results. However unlike hierarchical clustering, K-means lends itself naturally to an easily digestible visual display showing the centroid and silhouette for each cluster and enabling the user to see where in the cluster a certain gene is positioned. The BioGrid platform implements K-means clustering of gene expression data.

Hierarchical clustering

The main strength of hierarchical clustering lies in the fact that it does not require the user to predefine the number of clusters in the data. At the outset, each point is assigned its own cluster. The closest clusters are then merged and the process is repeated until we end up with a single cluster. Results of this analysis are commonly represented in the form of a dendrogram—a tree in which each branching is a single merge operation. A disadvantage of hierarchical clustering is the lack of a cut-off point, which determines the number of clusters. This can be defined manually after the clustering, but is impractical for large numbers of expression profiles. One way of dealing with this issue, which the BioGrid platform utilises, is to specify a tolerance constant that represents the minimum distance allowed between clusters.

2.4 Case study: Energy-related genes and protein complexes

For our investigations of the relationship between gene expression and protein interactions, we exploited the gene co-expression networks compiled by Stuart et al. [44]. In this study, Stuart and co-workers identified orthologous genes (i.e., genes in different organisms that have evolved from a single gene in an ancestral species) in humans, fruit flies, nematode worms, and baker's yeast on the basis of conserved protein sequences [46]. In total, they identified a set of 6 307 orthologous genes, representing 6 591 human genes, 5 180 worm genes, 5 802 fly genes, and 2 434 yeast genes. They then analyzed 3 182 DNA microarrays taken from these different organisms to identify pairs of orthologous genes whose expression profiles showed co-expression across multiple organisms and conditions. Compared to the conventional analysis of gene expression profiles from a single species, the use of orthologous genes across multiple species utilizes evolutionary conservation as a powerful criterion to identify genes that are functionally important among a set of co-expressed genes. Thus co-expression of a pair of orthologous genes over large evolutionary distances is indicative of a selective advantage to their co-expression, and hence the protein products of these genes are more likely involved in the same functional process. By applying the K-means algorithm described above, Stuart et al. were able to identify 12 gene clusters that each contained genes encoding proteins involved in specific cellular processes, e.g., signalling, cell cycle, and secretion.

The clusters of genes found by Stuart offer some important advantages over those from microarray experiments on single organisms, particularly with regard to studying protein interactions. Due to the evolutionary conservation that is implied, the observed orthologous genes are principally components of highly conserved biological processes, and it is probable that to increase efficiency, evolution has favored having some of these proteins coming together to form protein complexes, i.e., the genes of these expression clusters may be more likely to contain a significant proportion of interacting proteins. Such insights are less obvious in single species studies, where only correlated gene expression is implied. A cluster that particularly interested us from the data set was that containing a number of genes annotated as being involved in 'energy generation.'

3 Interaction space

The expression space is complemented by the interaction space. The former captures the activity of genes, the latter the interactions of proteins, which are important in providing a context to understand function.

3.1 Protein interactions are fundamental to understanding protein function

Protein-protein interactions are fundamental to most cellular processes [43]. The functions of proteins in biological systems are determined and mediated by the physical interactions they make with other molecules [53, 20]. The nature of these interactions ranges from short-lived interactions in signalling processes, to long-lived interactions between proteins of the skeletal components of a cell or organism. Protein-protein interactions occur in oligomers, between enzymes and substrates, in cell-cell contact through cell-adhesion molecules, and between antibodies and antigenes in the immune system. Despite the importance of protein interactions and technological advances in their detection, there is still a huge gap between the *circa* one million annotated proteins and around 50 000 documented interactions.

3.2 Experimental approaches for detecting protein interactions

There are a number of experimental approaches for detecting protein-protein interactions, for example:

- *Tandem affinity purification (TAP)*: A method for trapping and purifying a protein complex, based on the selective interaction of proteins with a protein that is attached to a solid support.

- *Co-immunoprecipitation (Co-IP)*: The use of a specific antibody to trap and purify a protein, plus the proteins it interacts with.
- *Phage display*: A technique that uses bacteriophages that have been genetically modified to express a new protein on their surface. Libraries of bacteriophages expressing many different proteins may be produced, and other proteins to which these bind may be purified and identified.

In a manner analogous to DNA microarrays, the techniques of TAP and Co-IP may be minaturized and used to detect multiple interactions simultaneously by printing sets of different peptides (peptide arrays) or antibodies (antibody arrays) onto slides. Another widely used technique for detecting protein-protein interactions is the 'yeast two-hybrid' (Y2H) method. To confirm an interaction of proteins A and B, the gene for protein A is fused with a gene encoding a DNA-binding domain for a specific reporter gene and the gene for protein B is fused with a gene encoding a transcription activation domain. Only if the proteins A and B interact, can the DNA-binding domain and the activation domain come together in order to initiate transcription of the reporter gene with its detectable product. An advantage of this approach is that interactions can be generated on a large-scale by breeding libraries of yeast cells with different genes fused to the DNA-binding and activation domains. Y2H data is collected and curated in databases such as the *Biomolecular Interaction Network Database* (BIND) [3] and the *Database of Interacting Proteins* (DIP) [57]. Although the Y2H method is used widely by the systems biology community, the method suffers from severe problems of false positives, i.e reporting that two proteins interact when in fact they do not *in vivo*. The reasons for this high false positive rate are most likely that either the reporter gene may be expressed independently of any interaction between proteins A and B, or that under normal physiological conditions, the two proteins are not expressed at the same time or location. There are estimates that between 50% to 80% of the interactions reported by Y2H are likely to be false positives. From these large sets of binary protein interactions may be generated maps, which represent the context and global structure of protein interaction networks.

3.3 Computational approaches for predicting protein interactions

Large-scale protein interaction maps from results of experimental methods [26, 36, 49, 52, 18, 19, 27, 16, 39] have increased our knowledge of protein function, extending 'functional context' to the network of interactions which span the proteome [23, 51, 15, 32]. Functional genomics fuels this new perspective, and has directed research towards *computational* methods of determining genome-scale protein interaction maps.

One group of computational methods uses the abundant genomic sequence data, and is based on the assumption that genomic proximity and gene fusion result from a selective pressure to genetically link proteins which physically interact [33, 8, 13]. However with the exception of polycistronic operons (where a set of neighbouring genes involved in a common process are under the control of a single operator and

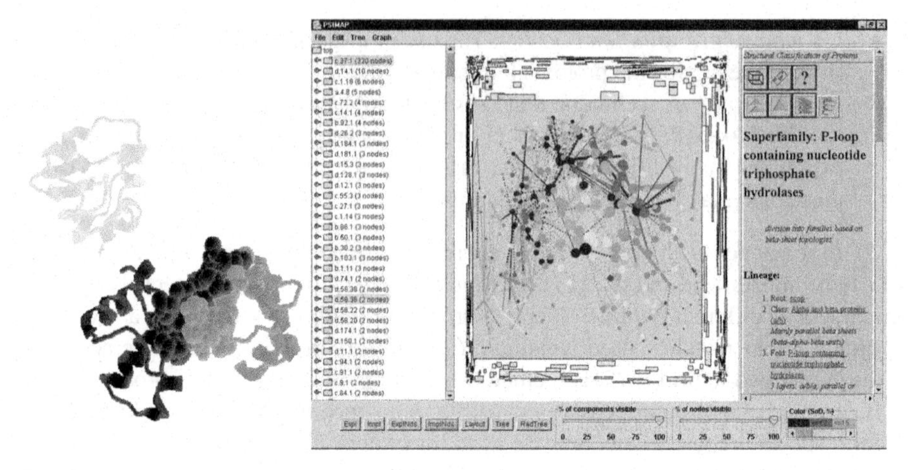

Fig. 2. The depicted structure contains three SCOP superfamilies: a 'winged helix DNA-binding domain' (a.4.5, medium gray, bottom right), an 'iron-dependent repressor protein, dimerization domain' (a.76.1, dark gray, bottom left), and a 'C-terminal domain of transcriptional repressors' (b.34.1, light gray, top). The possible interactions that could occur are: a.4.5–a.76.1, a.4.5–b.34.1, and a.76.1–b.34.1. However, in the structure, only the a.4.5–a.76.1 superfamily interaction is observed. The interacting residues of the two domains are depicted as small spheres. The *Protein Structure Interactome Map* (PSIMAP) determines these interactions for all multi-domain structures in the *Protein Data Bank* (PDB). On the right side of the figure, a screen shot of all such superfamily interactions in the PSIMAP database is depicted. It can be seen that the PSIMAP database contains a large number of independent components which contain only a few superfamilies. The main component in the middle of the figure contains 320 linked superfamilies. The most prominent superfamilies are the P-loop and immunoglobulin, which have both the most interaction partners and occur in the greatest number of different species.

thus expressed together), genomic proximity is only indicative of possible indirect functional associations between proteins [25], rather than direct physical interactions between the gene products.

A second group of methods, based on the assumption that protein-protein interactions are conserved across species, was originally applied to genomic comparisons [38]. Just as common function can be inferred between homologous proteins, 'homologous interaction' can be used to infer interaction between homologues of interacting proteins.

One approach to detect these interactions is with the *Protein Structure Interactome Map* (PSIMAP) algorithm. [37, 6, 7]. The PSIMAP algorithm finds interactions between protein domains in the *Protein Data Bank* (PDB) [4] using the domain definitions of the SCOP database[5]. As an example, consider Fig. 2. The depicted struc-

[5]The *Structural Classification of Proteins* (SCOP) database, created by manual inspection and abetted by a battery of automated methods, aims to provide a detailed and comprehensive description of the structural and evolutionary relationships between all proteins whose struc-

ture contains three SCOP domains: a winged helix DNA-binding domain (a.4.5.24, medium gray), an 'iron-dependent repressor protein, dimerization domain' (a.76.1.1, dark gray), and a 'C-terminal domain of transcriptional repressors' (b.34.1.2, light gray). The PSIMAP algorithm determines for a pair of domains whether there are at least five residue pairs within a five Å (Angstrom) distance. In this example, the PSIMAP algorithm determines that the DNA-binding domain (medium gray) interacts with the iron-dependent repressor protein (dark gray). The figure on the right highlights the atoms of the interacting residues as spheres, while the rest of the protein is shown as ribbons that follow the backbone of the proteins. The PSIMAP algorithm also determines that the DNA-binding domain and the transcriptional repressor domain are not interacting. The same holds for the two repressor domains. In both cases the distances are too great and the interactions are too few to constitute an interaction.

The PSIMAP algorithm finds such interactions for all multi-domain proteins in the PDB and the results are stored in the PSIMAP database, from which may be generated a map of interacting SCOP superfamilies. Domains of a SCOP superfamily are probably evolutionary related as evidenced by their common structure and function—despite having possibly low sequence similarity. Thus superfamily interactions are the appropriate level to study homologous interactions. The right side of Fig. 2 shows a screen shot of a global view of the map generated from the results in the PSIMAP database, depicting hundreds of superfamilies and their interactions.

The results in the PSIMAP database have been compared to experimentally determined domain interactions in yeast [37] and a correspondence of around 50% has been found. Given the high number of false positives in Y2H data [31], this result is very promising. The PSIMAP results have also been validated systematically at the sequence level using BLAST [35], and have been improved by the use of a statistical domain level representation of the known protein interactions [56, 10]. The PSIMAP database is also very comprehensive, being based upon 108 694 individual domain-domain interactions. This is an order of magnitude larger than the data available in the DIP database. The growth of the PDB also means that PSIMAP's coverage is increasing.

3.4 Case study: Energy-related genes and protein complexes

The PSIMAP results can be used to predict subunit and domain interactions in protein complexes. To this end, in [6] we studied two energy-related protein complexes: NADH:ubiquinone oxidoreductase and succinate dehydrogenase.

The protein complexes of the respiratory chain

The majority of molecular processes necessary for life are thermodynamically unfavorable, i.e., they require an input of energy to drive them, and thus need to be

ture is known. Within SCOP, the separate domains of proteins are identified and classified into a hierarchy.

coupled to a suitable thermodynamically favorable reaction in order to proceed. The most common source of energy used by cells to drive such reactions is the hydrolysis of adenosine triphosphate (ATP). Thus cells need a constant supply of ATP if they are to function and survive.

The most important mechanism for the synthesis of ATP is from the phosphorylation of adenosine diphosphate (ADP) by the enzyme ATP synthase. The source of energy for these endothermic reactions are electro-chemical gradients. These electrochemical gradients are generated from a series of redox reactions carried out by the protein complexes of the respiratory chain, which pump protons across a membrane leading to the establishment of a proton gradient and hence a proton motive force that may be used as an energy source to perform work. The protein complexes of the respiratory chain generate the electro-chemical gradient using the controlled reduction of molecular oxygen to water, and oxidation of sugars to carbon dioxide. This is a highly energetic reaction, however in the respiratory chain it is divided into a series of steps using electron transfer reactions between redox centers in the protein complexes so that the energy may be used to do useful work, i.e., to generate the electro-chemical gradient by pumping protons across a membrane.

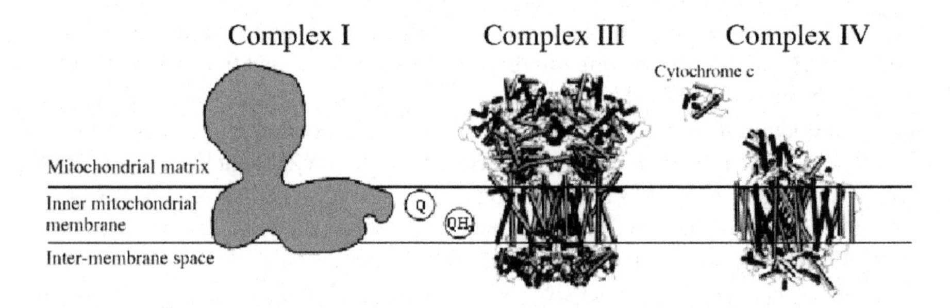

Fig. 3. A schematic of the three membrane-bound protein complexes of the respiratory chain in the inner mitochondrial membrane: NADH:ubiquinone oxidoreductase (complex I); ubiquinol:cytochrome c oxidoreductase (complex III) and cytochrome c oxidase (complex IV). Also shown are the two mobile electron carriers: ubiquinone (Q and QH2) and cytochrome c, that are responsible for shuttling electrons between the protein complexes.

The respiratory chain consists of three membrane-bound protein complexes: NADH:ubiquinone oxidoreductase (also known as complex I), ubiquinol:cytochrome c oxidoreductase (also known as complex III) and cytochrome-c oxidase (also known as complex IV). In addition, there are two mobile electron carriers: ubiquinone and cytochrome c (see Fig. 3). All the proteins of the respiratory chain are composed of multiple polypeptide units and incorporate a number of redox co-enzymes that are used to transport electrons, e.g., flavins, iron-sulphur centers, heme groups and copper ions. The passage of electrons from a molecule called NADH to molecular oxygen starts with complex I, where the electrons of NADH are passed via FMN and several iron-sulphur centers to ubiquinone, which is reduced to ubiquinol. The ubiquinol then dissociates from complex I and migrates through the mitochondrial membrane until it meets a molecule of complex III, at which point it is oxidized to ubiquinone and its electrons pass to complex III, which uses them to reduce cytochrome c. (An additional source of ubiquinol for complex III is succinate dehydrogenase (also known as complex II) of the citric acid cycle during the oxidation of succinate to fumarate.) Cytochrome c is oxidized by complex IV, which catalyses the transfer of electrons using copper ions and heme groups to their final destination of molecular oxygen. As complexes I, III and IV transfer electrons along their co-enzymes, it has the net effect of transporting protons from one side of the membrane to the other.

Since the components of the respiratory chain are large and predominantly hydrophobic, it has proven a major challenge to determine their structure by crystallography. Although structures are now known for complex II [59], complex III [58] and complex IV [48], as well as important subunits of ATP synthase [1], the structure of the relatively simple version of complex I found in E. coli with only 13 subunits has not yet been determined at atomic resolution. The human version of complex I has at least 45 proteins and the determination of its structure presents an even greater challenge. Thus alternative methods that may shed light on the structure, mechanism and evolution of these complexes are potentially useful. The work presented here is attempting to use the results of gene expression, homologous protein interactions and text analysis to identify and assemble the subunits of protein complexes involved in the generation of energy by cells, with a particular emphasis on complex I.

Recovering complex I and complex II

To evaluate whether protein interactions in complex I and complex II can be recovered using the superfamily interactions recorded in the PSIMAP database, we used the *Position Specific Iterative BLAST* (PSI-BLAST) application [2], to assign superfamilies defined in the SCOP database to known protein subunits of complex I and complex II. Thus known components of bovine complex I: 39 kDa subunit (Swiss-Prot:P34943), TYKY subunit (Swiss-Prot:P42028), and 75 kDa subunit (Swiss-Prot:P15690), were analysed and predicted to belong to the SCOP superfamilies '2Fe-2S ferredoxin-like' (d.15.4), 'nucleotide-binding domain' (c.4.1), '4Fe-4S ferredoxins' (d.58.1), and 'alpha-helical ferredoxin' (a.1.2), respectively. Furthermore, the two SCOP superfamilies, 'FMN linked oxidoreductase' (c.1.4) and 'FAD/NAD (P) binding domain' (c.3.1), are functionally significant to complex I. Protein components of complex II from nematodes: iron-sulfur subunit (Swiss-Prot:Q09545) and flavoprotein subunit (Swiss-Prot:Q09508), were found to map to '2Fe-2S ferredoxin-like' (d.15.4), 'alpha-helical ferredoxin' (a.1.2), 'succinate dehydrogenase/fumarate reductase flavoprotein C-terminal domain' (a.7.3), 'succinate dehydrogenase/fumarate reductase flavoprotein, catalytic domain' (d.168.1), and 'FAD/NAD(P)-binding domain' (c.3.1). Fig. 4 shows the induced subgraphs generated from the PSIMAP database using the predicted superfamilies of the complex I and complex II subunits. As a proof-of-principle, the known superfamily interactions

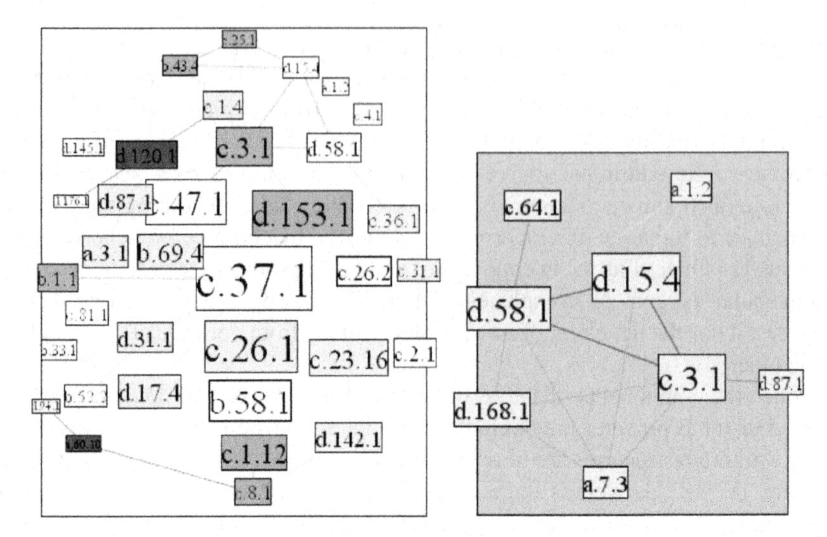

Fig. 4. Reconstructing complexes: For proteins of complex I and II, we assigned superfamilies to the protein domains. From the PSIMAP database, we obtained an induced subgraph of the superfamily interactions, i.e., only superfamilies of complex I and II, plus any superfamilies connecting these were selected. The superfamilies of complexes I and II are colored in light gray. The right side shows the induced subgraph for complex II—all complex II superfamilies interact directly with each other. The left shows complex I, whose structure is not yet solved—There are clear clusters of interacting superfamilies. In particular, the superfamily 'FAD/NAD(P)-binding domain' (c.3.1) appears to be part of complex I as it connects a number of complex I superfamilies.

of complex II, whose structure has been solved, are fully recovered. For complex I, whose structure is not yet solved, substantial numbers of interactions between predicted superfamilies of the subunits are predicted. Intermediate superfamilies connecting the predicted superfamiles of known subunits correspond to predicted superfamilies of complex I subunits that were not detected by the PSI-BLAST algorithm on the basis of sequence similarity.

4 Mapping expression data to interaction data

An assumption underlying the analysis of many microarray experiments is that if genes are co-expressed over a range of different conditions, then this is because they are being co-regulated by the cell, i.e., the protein products of the genes are involved in the same functional processes and are being controlled by a common set of transcription factors to ensure that they are all expressed together at the required time. A corollary to this is that if a set of proteins associate to form a protein complex, then it may be expected that the genes encoding these protein products would be co-regulated too—This is taken to its extreme in operons, where a single operator controls the expression of multiple genes as a polycistron. Thus an analysis of gene expression data to identify co-expressed genes over a range of conditions may identify putative components of protein complexes, as well as genes whose protein products are involved in similar functional processes [44, 12, 28, 24, 41].

Thus, can we relate the energy-related genes discussed in Section 2.4 with the electron-transport complexes discussed in the previous section? Before we can address this question, we need to link the expression and interaction space. This is not trivial, as the interaction space we presented is based on the limited structural data available in the PDB, however the structures of the majority of genes analyzed in the expression space will not have been determined. This knowledge gap may be bridged by comparing the sequences of proteins with known structural SCOP superfamilies to those proteins of unknown structure and assigning them a SCOP superfamily on the basis of sequence similarity. This approach is used by the SUPERFAMILY tool [22], which uses a library of hidden Markov models of domains of known structure from SCOP and provides structural (and hence implied functional) assignments to protein sequences at the superfamily level according to SCOP. This analysis has been carried out on all completely sequenced genomes, so the SUPERFAMILY database contains all the possible domain assignments for every gene of all completely sequenced genomes.

The BioGrid platform uses the SUPERFAMILY tool and database to link gene expression to protein interaction data by generating the induced interaction graphs for the genes of an expression cluster. For every gene within a gene expression cluster, we determine if there are domain assignments provided by the SUPERFAMILY tool. After all the domain assignments have been retrieved, the induced interaction graphs are generated using the PSIMAP database. The induced interaction network for a given set of superfamilies S is defined as the subgraph of the whole PSIMAP for the superfamilies S and the superfamilies on any shortest paths between any two superfamilies in S. Let us now consider the case study.

4.1 Case study: Energy-related genes and protein complexes

Table 2. Mapping between energy-related genes and SCOP superfamilies known to be part of complex II.

SCOP superfamily description	SCOP ID	Yeast gene names
FAD/NAD(P)-binding domain	c.3.1	YFL018C,YGR255C, YHR176W,YIL155C, YPL091W,YJL045W
Succinate dehydrogenase/fumarate reductase catalytic domain	d.168.1	YJL045W
Succinate dehydrogenase/fumarate reductase C-terminal domain	a.7.3	YJL045W
Alpha-helical ferredoxin domain	a.1.2	YLL041C
2Fe-2S ferredoxin-like domain	d.15.4	YLL041C

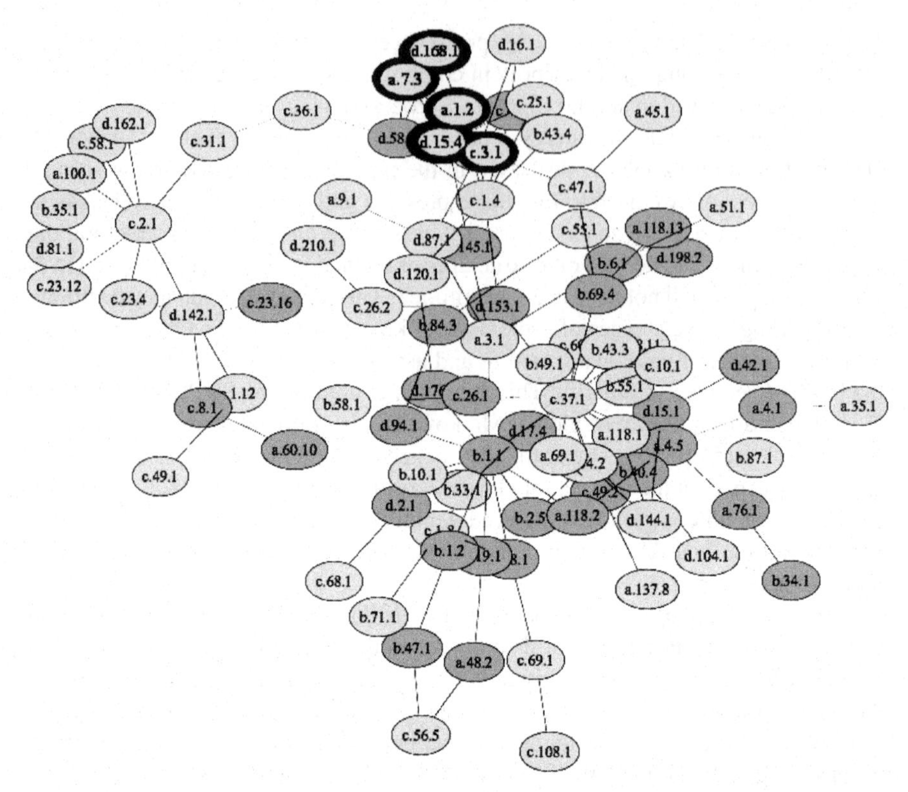

Fig. 5. Induced PSIMAP: The figure shows the superfamily interaction graph induced by the energy-related genes of an expression cluster from a set of microarray experiments [44]. Using the SUPERFAMILY tool, we determined the superfamilies assigned to the protein products of these genes. The graph shows the interaction network of these energy-related superfamilies (light gray) and any superfamilies, which are on any shortest path between two energy-related superfamilies (dark gray). In addition, five superfamilies known to be part of complex II occur in the cluster and have been circled in bold.

What protein domain interactions are predicted for the energy-related genes? Can we associate parts of energy-related protein complexes to these genes? To answer the first question, we determined the superfamilies for the energy-related genes and produced the induced interaction network as shown in Fig. 5. This shows the energy-related superfamilies in light gray and any superfamilies, which are on any shortest path between two energy-related superfamilies, in dark gray. Additionally, superfamilies known to be part of complex II have been circled in bold. These superfamilies can be linked back to the energy-related genes shown in Table 2. This example supports the link between the expression profiles of the energy-related gene cluster in the microarray data set and the physical interactions between subunits in the energy-related protein complexes.

5 Literature space

Any analysis in the expression and interaction space should be combined with an assessment of the relevant scientific literature. With the tremendous growth of literature this is not an easy task. PubMed, the main literature database referencing 6 000 000 abstracts, has grown by some 500 000 abstracts in 2003 alone. Due to this size, simple web-based text search of the literature is often not yielding the best results and a lot of important information remains buried in the masses of text. Text mining of biomedical literature aims to address this problem. There have been a number of approaches using literature databases such as PubMed to extract relationships such as protein interactions [5, 47]), pathways [17], and microarray data [45]. Mostly, these approaches aim to improve literature search by going beyond mere keyword search by providing natural language processing capabilities. While these approaches are successful in their remit, they do not mimic human information gathering.

Often scientists search the literature to discover new and relevant articles. They provide keywords and usually get back a possibly very long list of papers sorted by relevance. The search process can be broken down into three steps: First, a query may be pre-processed (e.g., keywords may be stemmed, synonyms may be included and general terms may be expanded (as done in PubMed)), second the search is carried out (this can range from a simple keyword search to refined concepts such as using document link structure as implemented in Google) and finally post-processing of relevant results (in most cases presentation of results as a list). While such lists are useful when looking up specific references, they are inadequate to get an overview over a large amount of literature and they do not provide a principled approach to discover new knowledge.

Our system, GoPubMed is based on mapping texts in paper abstracts to the Gene Ontology (GO) [6]. Gene Ontology is an increasingly important international effort to provide common annotation terms (a controlled vocabulary) for genomic and proteomic studies. The core of GO we are using is a term classification divided into three alternative directed acyclic graphs for molecular functions, biological processes, and cellular components. Two types of links are available: is_a and has_a. Multiple inheritance of subterms is possible.

To implement the literature space in the BioGrid platform, we provide a novel ontology-based literature search. GoPubMed allows one to submit keywords to PubMed and retrieve a given number of abstracts, which are then scanned for GO terms. The found terms are used to categorize articles and hence group related papers together. The hierarchical nature of the Gene Ontology gives the user the ability to quickly navigate from an overview to very detailed terms. Even with over 10 000 terms in the Gene Ontology, it takes a maximum of 16 terms to go from the root of the ontology to the deepest and most refined leaf concept. In particular GoPubMed works as follows:

[6]The Gene Ontology is a hierarchical vocabulary for molecular biology. (See `http://www.geneontology.org/`

Step 1 For each abstract, a collection of GO terms T is first found by using heuristics appropriate for the characteristic textual form of the GO terms.

Step 2 The minimal directed subgraph S is constructed that contains all the discovered terms T. The graph is constructed in XML to make presentation of the data in the HTML form easier. Because XML is a tree, not a graph, we clone and attach equivalent subtrees, which is required because of the multiple inheritance in GO.

Step 3 Statistics of each node are computed. For each node, we count all the paper links and discovered terms for the terms at the current node and the terms in the descendent nodes. The end result allows easily navigating to a subset of papers including a particular subcategory of terms (e.g., biosynthesis). Relative statistics can help to evaluate how important a particular process or function may be for the input query.

At the heart of GoPubMed is the problem of extracting Gene Ontology terms from free text. Finding exact terms in the literature is rarely possible and so GoPubMed employs a novel algorithm, which first tries to find short matching seed terms, which are then iteratively extended [9]. The subset of the Gene Ontology relevant to the retrieved papers and extracted terms is then used for exploration. Fig. 6 shows a screen shot of the system.

Example 1. For the example of energy-related genes and complex II superfamilies, we submitted the following SCOP superfamily names to GoPubMed, but limited the maximum number of retrieved abstracts to 40:

- 2Fe-2S ferredoxin-like (d.15.4).
- Alpha-helical ferredoxin (a.1.2).
- Succinate dehydrogenase/fumarate reductase C-terminal domain (a.7.3).
- Succinate dehydrogenase/fumarate reductase catalytic domain (d.168.1).
- FAD/NAD(P)-binding domain (c.3.1).

The relevant papers could be classified as shown in Table 3. All of them were classified as being concerned with electron transport and energy pathways.

6 Conclusion

In this paper we have given an overview of the BioGrid platform—an integrated platform for the analysis of gene expression and protein interaction data. The expression and interaction space is complemented by a literature space, which provides access to ontology-based literature searches. While the data and analysis of the individual spaces is well-understood and explored separately, there is little work on their integration to provide a holistic view of the underlying networks. The BioGrid platform addresses this problem. The example of energy-related genes and complexes illustrates the potential usefulness of this novel approach.

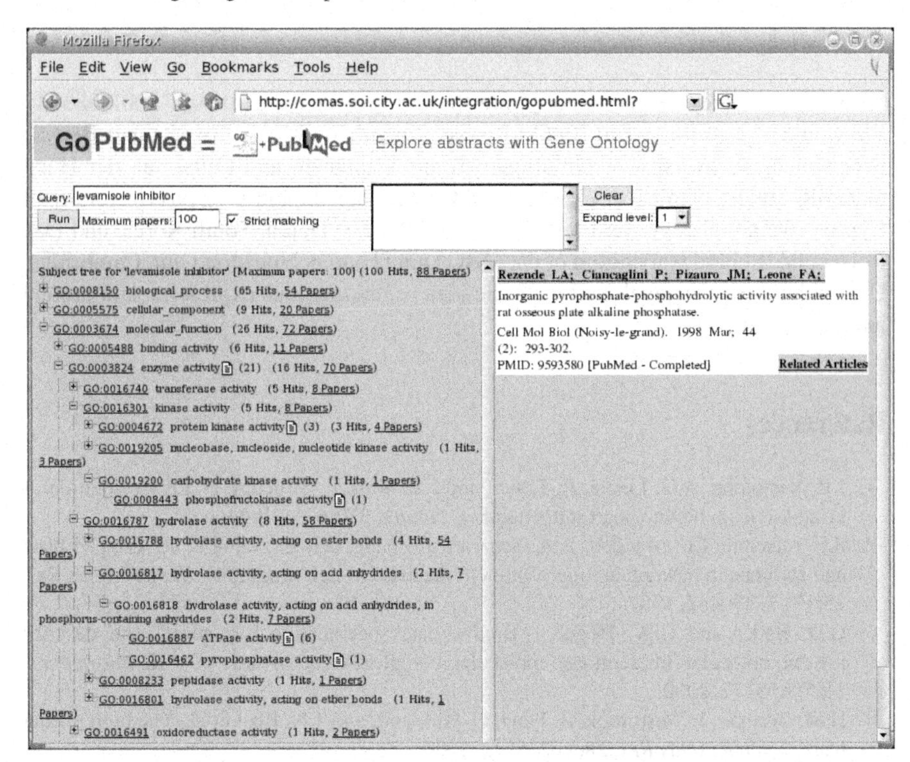

Fig. 6. User interface of GoPubMed. It displays the results for 'Levamisole inhibitor' limited to one hundred papers. A number of relevant enzyme activities are found.

Table 3. We submitted superfamily names to GoPubMed limiting the retrieval to 40 abstracts only. The table shows Gene Ontology terms in the process and function categories relevant to all five complex II superfamilies.

	electron transport	catabolism	vitamin	carbohydrate	coenz./prosthetic grp	energy pathways	oxidoreductase	transporter	binding
d.15.4	22	3	3	1	3	2	1	0	5
a.1.2	4	1	1	0	1	0	1	0	0
a.7.3	6	2	1	8	5	8	6	2	1
d.168.1	15	3	6	24	13	34	9	2	6
c.3.1	16	21	30	2	31	9	10	3	3

Acknowledgments

We wish to acknowledge support from the EU project BioGrid (IST-2002-38344). We would like to thank the BioGrid project members: Morris Swertz and Bert de Brock of the University of Groningen; Bram Stalknecht and Eelke van der Horst of ZooRobotics; Dimitrios Vogiatzis and George Papadopoulos of the University of Cyprus. We are also grateful to Jong Park of KAIST, Dejon, South Korea, and Dan Bolser and Richard Harrington of the MRC Dunn Human Nutrition Unit, Cambridge, UK, with whom we developed the basic idea of linking gene expression and protein interaction data.

References

1. J.P. Abrahams, A.G. Leslie, R. Lutter, and J.E. Walker. Structure at 2.8 a resolution of f1-atpase from bovine heart mitochondria. *Nature*, 370(621), 1994.
2. S.F. Altschul, T.L. Madden, A.A. Schaffer, J. Zhang, and Z. Zhang et al. Gapped blast and psi-blast: a new generation of protein database search programs. *Nucleic Acids Res*, 25(17):3389–402, 1997.
3. G.D. Bader and C.W. Hogue. Bind–a data specification for storing and describing biomolecular interactions, molecular complexes and pathways. *Bioinformatics*, 16(5):465–77, 2000.
4. H.M. Berman, J. Westbrook, Z. Feng, G. Gilliland, and T.N. Bhat et al. The protein data bank. *Nucleic Acids Res*, 28(1):235–42, 2000.
5. C. Blaschke, M.A. Andrade, C. Ouzounis, and A. Valencia. Automatic extraction of biological information from scientific text: protein-protein interaction. In *Proc. of the AAAI conf. on Intelligent Systems in Molecular Biology*, pages 60–7. AAAI, 1999.
6. D. Bolser, P. Dafas, R. Harrington, J. Park, and M. Schroeder. Visualisation and graph-theoretic analysis of the large-scale protein structural interactome network psimap. *BMC Bioinformatics*, 4(45), 2003.
7. P. Dafas, D. Bolser, J. Gomoluch, J. Park, and M. Schroeder. Fast and efficient computation of domain-domain interactions from known protein structures in the PDB. In H.W. Frisch, D. Frishman, V. Heun, and S. Kramer, editors, *Proceedings of German Conference on Bioinformatics*, pages 27–32, 2003.
8. T. Dandekar, B. Snel, M. Huynen, and P. Bork. Conservation of gene order: a fingerprint of proteins that physically interact. *Trends Biochem Sci*, 23(9):324–8, 1998.
9. R. Delfs, A. Kozlenkov, and M. Schroeder. Gopubmed: ontology-based literature search applied to gene ontology and pubmed. *Submitted*, 2004.
10. M. Deng, S. Mehta, F. Sun, and T. Chen. Inferring domain-domain interactions from protein-protein interactions. *Genome Res*, 12(10):1540–8, 2002.
11. J. DeRisi, V.R. Iyer, and P.O. Brown. Exploring the metabolic and genetic control of gene expression on a genomic scale. *Science*, 278:680–686, 1997.
12. M.B. Eisen, P.T. Spellman, P.O. Brown, and D. Botstein. Cluster analysis and display of genome-wide expression patterns. *Proc. Nat. Acad. Sci. USA*, 95(25):14863–14868, 1998.
13. A.J. Enright, I. Iliopoulos, N.C. Kyrpides, and C.A. Ouzounis. Protein interaction maps for complete genomes based on gene fusion events. *Nature*, 402(6757):86–90, 1999.

14. B.S. Everitt. *Graphical techniques for multivariate data.* Heinemann Educational Books, 1978.
15. M. Fellenberg, K. Albermann, A. Zollner, H.W. Mewes, and J. Hani. Integrative analysis of protein interaction data. In *Intelligent systems for molecular biology*, pages 152–61. AAAI Press, 2000.
16. M. Flajolet, G. Rotondo, L. Daviet, F. Bergametti, and G. Inchauspe et al. A genomic approach of the hepatitis c virus generates a protein interaction map. *Gene*, 242(1-2):369–79, 2000.
17. C. Friedman, P. Kra, H. Yu, M. Krauthammer, and A. Rzhetsky. Genies: a natural-language processing system for the extraction of molecular pathways from journal articles. In *Proceedings of the International Confernce on Intelligent Systems for Molecular Biology*, pages 574–82, 2001.
18. M. Fromont-Racine, A.E. Mayes, A. Brunet-Simon, J.C. Rain, and A. Colley et al. Genome-wide protein interaction screens reveal functional networks involving sm-like proteins. *Yeast*, 17(2):95–110, 2000.
19. M. Fromont-Racine, J.C. Rain, and P. Legrain. Toward a functional analysis of the yeast genome through exhaustive two-hybrid screens. *Nat Genet*, 16(3):277–82, 1997.
20. C.S. Goh, A.A. Bogan, M. Joachimiak, D. Walther, and F.E. Cohen. Co-evolution of proteins with their interaction partners. *JMB*, 299(2):283–93, 2000.
21. A.D. Gordon. *Classification.* Chapman and Hall, 1981.
22. J. Gough, K. Karplus, R. Hughey, and C. Chothia. Assignment of homology to genome sequences using a library of hidden markov models that represent all proteins of known structure. *J. Mol. Biol.*, 313(4):903–919, 2001.
23. L.H. Hartwell, J.J. Hopfield, S. Leibler, and A.W. Murray. From molecular to modular cell biology. *Nature*, 402:C47–C54, 1999.
24. T.R. Hughes, M.J. Marton, A.R. Jones, C.J. Roberts, and R. Stoughton et al. Functional discovery via a compendium of expression profiles. *Cell*, 102(109), 2000.
25. M. Huynen, B. Snel, W. Lathe, and P. Bork. Predicting protein function by genomic context: quantitative evaluation and qualitative inferences. *Genome Res*, 10(8):1204–10, 2000.
26. T. Ito, T. Chiba, R. Ozawa, M. Yoshida, and M. Hattori et al. A comprehensive two-hybrid analysis to explore the yeast protein interactome. *Proceedings of National Academy of Sciences USA*, 98(8):4569–4574, 2001.
27. T. Ito, K. Tashiro, S. Muta, R. Ozawa, and T. Chiba et al. Toward a protein-protein interaction map of the budding yeast: A comprehensive system to examine two-hybrid interactions in all possible combinations between the yeast proteins. *Proc Natl Acad Sci U S A*, 97(3):1143–7, 2000.
28. S.K. Kim, J. Lund, M. Kiraly, K. Duke, and M. Jiang et al. A gene expression map for caenorhabditis elegans. *Science*, 293(2087), 2001.
29. T. Kohonen. *Self-organising maps.* Springer–Verlag, 2nd edition edition, 1997.
30. J. Kruskal. The relationship between multidimensional scaling and clustering. In *Classification and clustering*. Academic Press, 1977.
31. J.H. Lakey and E.M. Raggett. Measuring protein-protein interactions. *Current opinion in structural biology*, 8(1):119–123, 1998.
32. M. Lappe, J. Park, O. Niggemann, and L. Holm. Generating protein interaction maps from incomplete data: application to fold assignment. *Bioinformatics*, 17(1):S149–56, 2001.
33. E.M. Marcotte, M. Pellegrini, H.L. Ng, D.W. Rice, and T.O. Yeates. Detecting protein function and protein-protein interactions from genome sequences. *Science*, 285(5428):751–3, 1999.

34. K.V. Mardia, J.T. Kent, and J.M. Bibby. *Multivariate analysis*. Academic Press, 1979.
35. L.R. Matthews, P. Vaglio, J. Reboul, H. Ge, and B.P. Davis et al. Identification of potential interaction networks using sequence-based searches for conserved protein-protein interactions or interologs. *Genome Res*, 11(12):2120–6, 2001.
36. S. McCraith, T. Holtzman, B. Moss, and S. Fields. Genome-wide analysis of vaccinia virus protein-protein interactions. *Proceedings of National Academy of Sciences USA*, 97(9):4879–4884, 2000.
37. J. Park, M. Lappe, and S.A. Teichmann. Mapping protein family interactions: intramolecular and intermolecular protein family interaction repertoires in the pdb and yeast. *J Mol Biol*, 307(3):929–38, 2001.
38. M. Pellegrini, E.M. Marcotte, M.J. Thompson, D. Eisenberg, and T.O. Yeates. Assigning protein functions by comparative genome analysis: protein phylogenetic profiles. *Proc Natl Acad Sci U S A*, 96(8):4285–8, 1999.
39. J.C. Rain, L. Selig, V. Battaglia, C. Reverdy, and S. Simon et al. The protein-protein interaction map of helicobacter pylori. *Nature*, 409(6817):211–5, 2001.
40. M. Schroeder, D. Gilbert, J. van Helden, and P. Noy. Approaches to visualisation in bioinformatics: from dendrograms to Space Explorer. *Information Sciences: An International Journal*, 139(1):19–57, 2001.
41. E. Segal, M. Shapira, A. Regev, D. Pe'er, and D. Botstein et al. Module networks: identifying regulatory modules and their condition-specific regulators from gene expression data. *Nat Genet.*, 34(166), 2003.
42. P.T. Spellman, G. Sherlock, M.Q. Zhang, V.R. Iyer, and K. Anders et al. Comprehensive identification of cell cycle regulated genes of the yeast saccharomyces cerevisiae by microarray hybridization. *Molecular Biology of the Cell*, 9(12):3273–3297, 1998.
43. E. Sprinzak and H. Margalit. Correlated sequence-signatures as markers of protein-protein interaction. *JMol Bio*, (4):681–692, 2001.
44. J.M. Stuart, E. Segal, D. Koller, and S.K. Kim. A gene-coexpression network for global discovery of conserved genetic modules. *Science*, 302(249), 2003.
45. L. Tanabe, U. Scherf, L.H. Smith, J.K. Lee, and L. Hunter et al. Medminer: internettext-mining tool for biomedical information, with application to gene expression profiling. *BioTechniques*, 27(6):1210–4,1216–7, 1999.
46. R.L. Tatusov, E.V. Koonin, and D.J. Lipman. A genomic perspective on protein families. *Science*, pages 631–637, 1997.
47. J. Thomas, D. Milward, C. Ouzounis, S. Pulman, and M. Carroll. Automatic extraction of protein interactions from scientific abstracts. In *Proc. of the Pacific Symp. on Biocomputing*, pages 538–49, 2002.
48. T. Tsukihara, H. Aoyama, E. Yamashita, T. Tomizaki, and H. Yamaguchi et al. The whole structure of the 13-subunit oxidized cytochrome c oxidase at 2.8 a. *Science*, 272(1136), 1996.
49. P. Uetz, L. Giot, G. Cagney, T.A. Mansfield, and R.S. Judson et al. A comprehensive analysis of protein-protein interactions in saccharomyces cerevisiae. *Nature*, 403(6770):623–7, 2000.
50. J. van Ryzin, editor. *Classification and clustering*. Academic Press, 1977.
51. M. Vidal. A biological atlas of functional maps. *Cell*, 104(3):333–340, 2001.
52. A.J. Walhout, R. Sordella, X. Lu, J.L. Hartley, and G.F. Temple et al. Protein interaction mapping in c. elegans using proteins involved in vulval development. *Science*, 0(5450):116–121, 1999.
53. A.J. Walhout and M. Vidal. Protein interaction maps for model organisms. *Nature Reviews*, 2(1):55–62, 2001.

54. P. Wang, editor. *Graphical representations of multivariate data.* Academic Press, 1978.
55. A. Webb. *Statistical pattern recognition.* Arnold, 1999.
56. J. Wojcik and V. Schachter. Protein-protein interaction map inference using interacting domain profile pairs. *Bioinformatics*, 17(1):S296–305, 2001.
57. I. Xenarios, L. Salwinski, X.J. Duan, P. Higney, and S.M. Kim et al. Dip: the database of interacting proteins. *Nucleic Acids Research*, 28(1):289–291, 2000.
58. D. Xia, C.A. Yu, H. Kim, J.Z. Xia, and A.M. Kachurin et al. Crystal structure of the cytochrome bc1 complex from bovine heart mitochondria. *Science*, 277(60), 1997.
59. V. Yankovskaya, R. Horsefield, S. Tornroth, C. Luna-Chavez, and H. Miyoshi et al. Architecture of succinate dehydrogenase and reactive oxygen species generation. *Science*, 299(700), 2003.

Ontologies in Bioinformatics and Systems Biology

Patrick Lambrix

Department of Computer and Information Science
Linköpings universitet, Sweden.
E-mail: patla@ida.liu.se

Summary. Ontologies are being used in bioinformatics and systems biology, among others, for communication between people and organizations, as the basis for interoperability between systems, and as query models and indexes to repositories of information. In this chapter we give a background of the area and provide a state of the art overview. We present different possible definitions of ontology, examples of bio-ontologies and their use, formalisms that can be used to represent ontologies as well as tools that support the different stages in the life cycle of an ontology.

1 Introduction

Intuitively, ontologies can be seen as defining the basic terms and relations of a domain of interest, as well as the rules for combining these terms and relations. Ontologies are being used nowadays in many areas, including bioinformatics and systems biology, among others, for communication between people and organizations, as the basis for interoperability between systems, and as query models and indexes to repositories of information. Although ontologies have been around for a while, it is only during the last five years that the creation and use of ontologies in bioinformatics and systems biology have emerged as important topics. The number of researchers working on methods and tools for supporting ontology engineering is constantly growing and more and more researchers and companies use ontologies in their daily work. The work on ontologies is now also recognized as essential in some of the grand challenges of genomics research [7].

In this chapter we give a background of the area and provide a state of the art overview. We present different possible definitions of ontology (Section 2), talk briefly about ontologies in bioinformatics and systems biology (bio-ontologies) (Section 3) and show their use (Section 4). Then, we discuss an issue that is important for the use of an ontology, namely its representation language and its level of formality. We show some of the consequences of the different possible choices (Section 5). Finally, we pay attention to ontology engineering tools. These are tools that support the different stages in the life cycle of an ontology such as its creation and maintenance.

W. Dubitzky and F. Azuaje (eds.), Artificial Intelligence Methods and Tools for Systems Biology, 129–145.

We describe the main categories and some of the most well-known systems in this field, and show the results of evaluations of these tools (section 6).

2 Ontologies

The word 'ontology' has an original meaning in philosophy where it is the study of being, the study of existence. Within the field of artificial intelligence we find several definitions. One of the earlier is: An ontology defines the basic terms and relations comprising the vocabulary of a topic area, as well as the rules for combining terms and relations to define extensions to the vocabulary [34]. A definition that has been referenced a lot is the following by Gruber [19]: An ontology is an explicit specification of a conceptualization. A conceptualization is an abstract, simplified view of the world that we want to represent. According to this view, an ontology contains definitions and axioms. Definitions associate the names of entities in the domain with human-readable text describing what they are meant to denote. Formal axioms constrain the interpretations and well-formed use of these names of entities. This definition of ontology has been adapted by others. For instance, some definitions require the specification to be formal and some definitions require the conceptualization to be shared among many users. Still others have provided definitions based on the way they have used the ontologies (e.g., as a basis for the construction of knowledge bases).

Guarino and Giaretta [20] try to bring some clarity in the terminology and discuss the different interpretations they have found in the literature: Ontology as (1) a philosophical discipline, (2) an informal conceptual system, (3) a formal semantic account, (4) a specification of a 'conceptualization', (5) a representation of conceptual system via a logical theory, (6) the vocabulary used by a logical theory, and (7) a (meta-level) specification of a logical theory. The first interpretation is very different from the others as ontology is seen as a discipline. The other interpretations refer to ontologies as particular objects. In interpretations two and three an ontology is seen as a semantic entity. The ontology is a formal or informal conceptual system where a conceptualization is defined as an intensional semantic structure which encodes the implicit rules constraining the structure of a piece of reality. Interpretations five, six and seven on the other hand consider an ontology as a syntactic entity. In interpretation five the ontology is a logical theory, in interpretation six the vocabulary of such a theory and in interpretation seven the ontology defines the primitives used in the logical theory. Guarino and Giaretta state that when 'conceptualization' in interpretation four is interpreted as vocabulary then interpretation four collapses into interpretation five. Based on these interpretations (two to seven) Guarino and Giaretta suggest then two senses for the word 'ontology': (i) as synonym of conceptualization and (ii) a logical theory which gives an explicit, partial account of a conceptualization. This means that ontology is used as a rather high-level concept and that, when more precise definitions are needed, we can specify the use of the word 'ontology' by referring to the right interpretation.

As we have shown, giving a definition for ontology is not an easy task. This is complicated even more by the fact that ontologies differ with respect to the kind of information that they represent (e.g., Section 5) as well as the way they are used (e.g., Section 4). Also, work concerning ontologies is at different stages of development and maturity in different research communities. Therefore, giving one final definition of ontology at this point in time may not be possible. However, for reasons of clarity, reuse and reproducibility, researchers should clearly specify in their work what information is represented in their ontologies and in which representation framework, and how the ontologies are used. For a larger overview and discussion of definitions we refer to [18].

3 Bio-ontologies

Within the bioinformatics and systems biology fields ontologies have been around for a while and there exist many kinds of ontologies. Many of the model organism databases such as Flybase and Mouse Genome database can be seen as simple ontologies. Further, there are ontologies focusing on things such as protein functions, organism development, anatomy and pathways. Some examples are the thesaurus MeSH (Medical Subject Headings) [32], the MGED (Microarray Gene Expression Data) [33] ontology which aims to provide standard terms for the annotation of microarray data, STAR/mmCIF [50] which is an ontology for macromolecular structure, the TAMBIS ontology [3] which combines knowledge from several databases such as Swiss-Prot and PROSITE, and the ontology for biological function developed for the EcoCyc DB [15]. The field has matured enough to start talking about standards. An example of this is the organization of the first conference on Standards and Ontologies for Functional Genomics (SOFG) in 2002 and the development of the SOFG resource on ontologies [48].

The use of ontologies in bioinformatics has grown drastically since database builders concerned with developing systems for different (model) organisms joined to create the Gene Ontology Consortium (GO, [8, 17]) in 1998. For a complete list of member organizations we refer to the Gene Ontology Consortium home page [17]. The goal of GO is to produce a structured, precisely defined, common and dynamic controlled vocabulary that describes the roles of genes and proteins in all organisms [8]. Currently, there are three independent ontologies publicly available over the Internet: biological process, molecular function and cellular component. The GO ontologies have nowadays become a de facto standard and are used by many databases containing information about genes and proteins for annotation. In Fig. 1 we show a small part of a GO ontology. The top level concept in this part is $defense\ response$. Indentation represents an is-a (generalization-specialization) relationship. Thus, $hypersensitive\ response$ is a $defense\ response$ and $antigenpresentation$ is an $immune\ response$ which, in its turn, is a $defenseresponse$.

Recently, Open Biological Ontologies (OBO, [40]) was started as an umbrella web address for ontologies for use within the genomics and proteomics domains. The member ontologies are required to be open, to be written in a common syntax, to be orthogonal to each other, to share a unique identifier space and to include textual definitions. Many bio-ontologies are already available via OBO.

defense response ; GO:0006952
  hypersensitive response ; GO:0009626
  immune response ; GO:0006955
    acute-phase response ; GO:0006953
    antigen presentation ; GO:0019882
      antigen presentation, endogenous antigen; GO:0019883
      antigen presentation, exogenous antigen; GO:0019884

Fig. 1. Part of a GO ontology describing defense response.

4 Use of ontologies

Ontologies are used in various ways and this may have an influence on what information is represented in the ontologies. For instance, there are domain-oriented ontologies, task-oriented ontologies and generic ontologies. Most of the bio-ontologies are a mixture of these types of ontologies [49]. Jasper and Uschold [24] propose a classification of uses of ontologies (that was extended in [49]). They define the following scenarios: neutral authoring, ontology as specification, common access to information and ontology-based search. We describe the types and give examples for bio-ontologies.

- *Neutral authoring.* Within the neutral-authoring scenarios application-neutral ontologies are developed in a single language. The knowledge is then converted into a different form (e.g., a knowledge base) for use in multiple target applications. The benefits include reuse and portability of knowledge across platforms (e.g., the knowledge in the GO ontologies is used in different data banks) and improved maintainability (changes in the knowledge need only be made in one place).
- *Specification.* In the ontology-as-specification scenarios an ontology is used as a basis for software development. An example is the definition of a database schema or of a vocabulary for annotation. More and more information sources annotate their entries with terms from an ontology. Search engines can take advantage of this annotation as it gives extra information. Also, it allows for a simple form of integration of information sources as well as knowledge discovery as there is a good chance that there exist relationships between entries in different sources that are annotated with the same term. Benefits of the ontology-as-specification scenarios include documentation, maintenance, reliability, sharing and knowledge reuse.
- *Common information access.* The third kind of scenarios deals with common access to information. Ontologies are used to make information that needs to be

shared intelligible to multiple applications or humans. One example of such a scenario is the case where the ontology is used to promote common and consistent understanding among people and organizations. The ontology is shared among different people who all reference it in their work. The ontology may become a community reference. Several of the current bio-ontologies developers state this as a goal. The ontology can also be used as an interchange format between different applications. This may improve interoperability.
- *Search.* Finally, in the ontology-based search scenarios the ontology is used for querying over information sources with as aim to improve the quality of the answers (e.g., better precision and recall) and to reduce the time spent searching (e.g., [28]). For instance, the ontology may be used as an index to the information in the information sources. A user can browse the ontology and use the terms in the ontology as query terms. The ontology may also be used for query refining and expansion by moving up and down in the hierarchy of concepts. For instance, when a user searches in a database for 'immune response' and gets only very few results, she may decide to query with a more general term to find more answers. The ontology can be used to find these more general terms, in this case, for instance 'defense response'.

Some of the more advanced information sources use logical inferencing (see Section 5). In this case the ontology may be used to guide this reasoning. For instance, while looking for defense responses also all entries on immune responses will be returned as the ontology states that immune responses are defense responses. Another area where ontologies are proving to be useful is the area of data mining and knowledge discovery. For instance, in [1] a web tool for finding associations of GO terms with groups of genes is presented. Ontologies can also be used for clearly separating domain knowledge from application-based knowledge. Further, using ontologies forces system developers to make domain assumptions explicit. The ontologies can also be used for validation which leads to more reliable software.

5 Knowledge representation for ontologies

Ontologies differ regarding the kind of information they can represent. From a knowledge representation point of view ontologies can have the following components (e.g., [49]).

- *Concepts* represent sets or classes of entities in a domain. For instance, *immune response* represents all the things that are immune responses. The concepts may be organized in taxonomies, often based on the is-a relation or the part-of relation.
- *Instances* represent the actual entities. They are, however, often not represented in ontologies.
- Further, there are many types of *relations*. For instance, one type is the group of taxonomic relations such as the specialization relationships (e.g., *immune*

response is-a *defense response*) and the partitive relationships (e.g., *cell* hasComponent *nucleus*). Another type are the associative relationships such as nominative relationships that describe the name of concepts (e.g., *protein* hasName *proteinname*) and locative relationships that describe the location of a concept with respect to another (e.g., *chromosome* hasSubcellularLocation *nucleus*).

- Finally, *axioms* represent facts that are always true in the topic area of the ontology. These can be such things as domain restrictions (e.g., the origin of a *protein* is always of the type *gene coding origin type*), cardinality restrictions (e.g., each *protein* has at least one *source*), or disjointness restrictions (e.g., a *helix* can never be a *sheet* and vice versa).

Depending on which of the components are represented and the kind of information that can be represented, we can distinguish between different kinds of ontologies. A simple type of ontology is the *controlled vocabulary*. Controlled vocabularies are essentially lists of concepts. When these concepts are organized into a generalization-specialization hierarchy, we obtain a *taxonomy*. A slightly more complex kind of ontology is the *thesaurus*. In this case the concepts are organized in a graph. The arcs in the graph represent a fixed set of relations, such as synonym, narrower term, broader term, similar term. The *data models* allow for defining a hierarchy of classes (concepts), attributes (properties of the entities belonging to the classes, functional relations), relations and a limited form of axioms. There are also the *knowledge bases* which are often based on a logic. They can represent all types of components and provide reasoning services such as checking the consequences of the statements in the ontology and building the generalization-specialization hierarchy.

An ontology and its components can be represented in a spectrum of representation formalisms ranging from very informal to strictly formal. This would include natural language, limited and structured forms of natural language, formally defined languages and logics with formal semantics [24]. The choice of which formalism to use depends on the characteristics of the ontology as well as on its intended use. For instance, as controlled vocabularies are essentially lists of words, natural language may be a suitable representation formalism and the machinery of a formal logic may not be needed. Taxonomies are often represented in a formalism that introduces some structure. An example of this is XML (Extensible Markup Language). Due to XML's portability and widespread acceptance, a number of XML-based formalisms have been developed. An example is SBML (Systems Biology Markup Language) [23, 46] which is an XML-based format for representing biochemical reaction networks. Formalisms such as Entity-Relationship diagrams or UML (Unified Modeling Language) can be used for the data models. Also, database schemata can be considered as a possible representation of such ontologies. Frame systems define a number of language constructs (frames, slots, facets, values). Frames represent classes or instances. Classes are concepts in the domain. Slots describe properties or attributes of classes. Facets describe properties of slots and axioms define additional constraints.

Frame systems impose restrictions on how the language constructs are used or combined to define a class.

The most expressive representation formalisms in use for ontologies are the logics. Logics are formal languages with well-defined syntax, semantics and inference mechanisms. Logics provide support in ontology design, ontology aligning and merging and ontology deployment. Regarding ontology design, for instance, they allow for checking of concept and ontology satisfiability and for computing implied relationships. Regarding ontology aligning and merging they support the assertion of inter-ontology relationships, computing an integrated concept hierarchy, and consistency checking. In the ontology deployment phase, they support determining, for instance, whether a set of facts is consistent with respect to an ontology and whether instances belong to ontology concepts.

An interesting family of logics for representing ontologies are description logics (e.g., [2]). Description logics are knowledge representation languages tailored for expressing knowledge about concepts and concept hierarchies. As they are logics they have a well-defined semantics. They are considered an important formalism unifying and giving a logical basis to the well known traditions of frame-based systems, semantic networks and KL-ONE-like languages, object-oriented representations, semantic data models, and type systems. The basic building blocks are concepts, roles and individuals. Concepts describe the common properties of a collection of individuals. Roles represent binary relations. Each description logic defines also a number of language constructs (such as intersection, union, role quantification, etc.) that can be used to define new concepts and roles. The main reasoning tasks are classification and satisfiability, subsumption and instance checking. Subsumption represents the is-a (generalization-specialization) relation. Classification is the computation of a concept hierarchy based on subsumption. Several knowledge representation systems have been built using these languages and have been used in a variety of applications. They are now also seen as a base technology for the semantic web where content is labelled with semantic annotations and these annotations are used for automated retrieval and composition of web content.

One of the languages within this family, DAML+OIL [12], was previously recommended by the BioOntology Consortium as its choice of ontology representation language. Its successor, OWL [42], is now one of the languages that may be used for representing OBO ontologies. Both languages are XML-based and are closely related to the SHIQ(D) and SHOQ(D) description logics [22]. In Fig. 2 we give examples of a subset of the concept constructors in the DAML+OIL language. Using the IntersectionOf constructor new concepts can be defined as the intersection of previously defined concepts. For instance, $signal\ transducer\ activity \sqcap binding$ represents all things that are both a $signal\ transducer\ activity$ and a $binding$. The other boolean operations are handled by the UnionOf and ComplementOf constructors. For instance, $helix \sqcup sheet$ represents all things that are helices or sheets, while the things that are not helices are represented by $\neg\ helix$. The language also allows the use of existential and universal quantifiers and number restrictions. For instance, $\exists$ hasOrigin. $mitochondrion$ represents all things that originate from a $mitochondrion$, while $\forall$ hasOrigin. $gene\ coding\ origin\ type$ represents all things that have only origins belonging to the type *gene coding origin type*. The things having at least one source are represented by $\geq$ 1 hasSource. For the complete syntax and the semantics of DAML+OIL and OWL we refer to [12] and [42], respectively.

IntersectionOf	$C \sqcap D$	*signal transducer activity* $\sqcap$ *binding*
UnionOf	$C \sqcup D$	*helix* $\sqcup$ *sheet*
ComplementOf	$\neg$ C	$\neg$ *helix*
ToClass	$\forall$ R.C	$\forall$ hasOrigin. *gene coding origin type*
HasClass	$\exists$ R.C	$\exists$ hasOrigin. *mitochondrion*
MinCardinality	$\geq$ n R	$\geq$ 1 hasSource

Fig. 2. Some DAML+OIL concept constructors.

In general, the more formal the representation language, the less ambiguity in the ontology. The ontologies defined in an informal way rely on shared human consensus. However, there is no guarantee that there are no hidden ambiguities or hidden assumptions. This is more easy to check for formal languages. Formal languages are also more likely to implement correct functionality. Further, the chance for interoperation is higher. There is no common ground for interoperability based on the informal languages. In the informal languages the ontology content is also hard-wired in the application. This is not the case for the formal languages as they have a well-defined semantics. However, building ontologies using formal languages is not an easy task.

In practice, in bioinformatics and systems biology ontologies such as the GO ontologies, have started out as controlled vocabularies. This allowed the ontology builders, which were domain experts, but not necessarily experts in knowledge representation, to focus on the gathering of knowledge and the agreeing upon definitions. More advanced representation and functionality was a secondary requirement and was left as future work. However, some of the bio-ontologies have reached a high level of maturity and stability regarding the ontology engineering process and their developers have now started investigating how the usefulness of the ontologies can be augmented using more advanced representation formalisms and added functionality.

6 Ontology tools

In the same way as there exist many tools in software engineering that provide support for the different software development phases, there are now also tools that provide support for the different phases of the ontology engineering process. Based on the tasks and processes that are supported, the ontology tools can be grouped in the following clusters [9]. *Ontology development tools*, such as Protégé-2000, Chimaera, OilEd, Ontolingua, WebOnto, Ontosaurus and KADS, are used for building new ontologies. These tools usually support editing, browsing, documentation, export and import from different formats, views, libraries and they may have attached inference engines.

Ontology merge and integration tools (e.g., Protégé-2000 with PROMPT, Chimaera, SAMBO) support users in merging or integrating ontologies in the same domain. Within an area there are always a number of ontologies, each with their own focus. For instance, in bioinformatics ontologies may cover different aspects in molecular biology such as molecular function and cell signaling. Many of these ontologies contain overlapping information. For instance, a protein can be involved in both cell signaling and other biological processes. In applications using ontologies it is therefore of interest to be able to use multiple ontologies. However, to obtain the best results, we need to know how the ontologies overlap and align them (define the relationships between the ontologies) or merge them (create a new ontology containing the knowledge included in the original ontologies). Another reason for merging ontologies is that it allows for the creation of ontologies that can later be composed into larger ontologies. Also, companies may want to use de facto standard ontologies and merge them with company-specific ontologies.

Ontology evaluation tools, such as OntoAnalyzer, OntoClean and ONE-T, support ensuring a certain level of quality for the ontologies. A kind of tools that may become more important for bioinformatics and systems biology are the *ontology-based annotation tools*, which allow users to insert ontology-based mark-up in web pages. Further, there are also *ontology learning tools* that derive ontologies from natural language texts and *ontology storage and querying tools*. In this section we briefly describe the tools that have already been used in evaluations in the field of bioinformatics and present some evaluations of ontology tools.

6.1 Tools

Protégé-2000 [39, 44] is a software for creating, editing and browsing ontologies developed by Stanford Medical Informatics. The design and development of Protégé-2000 has been driven by two goals: to be compatible with other systems for knowledge representation and to be an easy to use and configurable tool for knowledge extraction. Protégé-2000 is available as free software, is Java-based and should be installed locally. It has many plug-ins that extend the functionality of the system. For instance, the plug-in PROMPT [43] supports merging and aligning of ontologies [36].

The Protégé knowledge model is frame based. An ontology consists of classes, slots, facets and axioms [35]. Protégé-2000 provides ontology editing functionality on different levels. Classes, attributes and instances can be created, added, deleted, viewed and searched for. Super-classes can be added, deleted and replaced. Further, it is possible to query the ontology. The Protégé-2000 user interface contains several tabbed panes where in each pane relevant information about a selected ontology component is shown. The ontologies are shown using a tree structure. Plug-ins are provided for querying based on F-Logic [25], merging and annotation of the ontologies with WordNet [51]. The information exchange mechanisms are based on a restricted version of Open Knowledge Base Connectivity (OKBC, [5]), an application programming interface for frame-based knowledge representation systems.

Protégé-2000 supports the import of text files, database tables and RDF (Resource Description Framework) files. The user can save her work in three different formats: as standard text files, in a JDBC database and in RDF format. Plug-ins can be used to support other formats such as XML. The user can customize Protégé-2000 in the sense that she can choose between different layouts and she can choose which tabbed panes are shown in the user interface and in which order.

Chimaera [31, 30, 6] is developed by the Knowledge Systems Laboratory at Stanford University and aims to provide assistance to users for browsing, creating, editing, merging and diagnosing of ontologies. It is available over the Internet and requires a relatively fast connection to be able to work efficiently. It was built on top of the Ontolingua Distributed Collaborative Ontology Environment. The initial goal was to develop a tool that could give substantial assistance for the task of merging knowledge bases produced by different users for different purposes with different assumptions and different vocabularies. Later the goals of supporting testing and diagnosing ontologies arose as well. The user interacts with Chimaera through a browser such as Netscape or Microsoft Internet Explorer.

Chimaera's knowledge model is frame based. An ontology consists of classes and slots. Chimaera provides about seventy commands in the user interface, thereby enabling taxonomy and slot editing. The applicable commands at each point in time are made available by the interface. In addition to editing, some of the commands are related to ontology merging. There are also commands related to diagnosis that, among others, check for incompleteness, cycles and value-type mismatches. All components of the ontology are shown in one window using a tree structure. As the current user interface is not a general-purpose editing environment, non-slot individuals and facets are not displayed [31]. The information exchange mechanisms are based on OKBC.

Chimaera accepts as input 15 different formats among which Protégé files and OKBC. The user can create knowledge bases of types ATP, CLOS, Ontolingua and Tuple KB. Chimaera has many different alternatives with respect to customization such as choosing default analysis tools, autosave and translating uploaded ontologies automatically to OKBC. The focus is on customization of the behavior of the tool.

OilEd [4, 45, 41] is a graphical tool for creating and editing OIL ontologies developed at the University of Manchester. The tool should be installed locally. One of the goals of OilEd is to show the use of the DAML+OIL language.

The knowledge model for OilEd is based on description logics. In contrast to frame systems, OilEd allows for arbitrary boolean combinations of classes. It also allows several types of constraints such as value-type and cardinality restrictions. OilEd provides creation and editing functionality for classes, relations, instances and axioms. The OilEd user interface has been strongly influenced by Protégé-2000. The ontologies are shown using a tree structure. OilEd uses the FaCT system [21, 16], a description logic system, for reasoning services such as checking the consequences of the statements in the ontology, classification and consistency checking.

OilEd supports import from Simple RDFS, DAML+OIL, SHIQ, HTML, Dotty and DIG and exports DAML+OIL. It allows a limited amount of customizations pertaining partly to the look of the tool and partly to the reasoning capabilities.

DAG-Edit [11] is open source software implemented in Java and should be installed locally. The tool offers a graphical user interface to browse, search and edit Gene Ontology files or other ontologies based on the directed acyclic graph (DAG) data structure. The relationships that are supported are is-a and part-of. DAG-Edit has its main focus on browsing and editing GO. Functionality is provided for the creation and deletion of concepts, adding synonyms, adding database references, and merging concepts. DAG-Edit shows the components of the ontology in one window. The ontologies are shown using a tree structure. The save formats in DAG-Edit are GO and OBO formats. DAG-Edit supports configuration via the Configuration Manager Plugin. The user can choose formats and what plug-ins should be shown.

SAMBO [29] is an ontology merge system developed at Linköpings universitet. It helps a user to merge two DAML+OIL ontologies into a new DAML+OIL ontology. Similarly to PROMPT and Chimaera, the system generates suggestions for merging concepts or relationships and for creating is-a relationships. The user can accept or reject these suggestions and can also add own suggestions. SAMBO performs the actual merging and computes the consequences of the merge. The SAMBO system uses the FaCT reasoner to provide a number of reasoning services such as consistency, satisfiability and equivalence checking.

6.2 Evaluations of ontology tools

Currently, only few evaluations of ontology tools using bio-ontologies have been performed. In [27] and [26] different tools were evaluated and compared as ontology tools. In [52] one tool was assessed for its use in developing and maintaining GO ontologies. Each evaluation had a different purpose and although GO has been used in all evaluations, there is no benchmark test suite yet that can easily be used by future evaluations. In this section we briefly describe the results of these three evaluations and then also give a short overview of evaluations of ontology tools in other areas.

In [27] Protégé-2000, Chimaera, OilEd and DAG-Edit were evaluated as ontology development tools using GO ontologies as test ontologies. The different systems were in different phases of development and further development was ongoing for most of them. The systems were evaluated according to the following criteria:

- *Availability*. How is the tool used: local installation or via the web?
- *Functionality*. What functionality does the tool provide?
- *Multiple inheritance*. Is multiple inheritance supported? How is it visualized in the tools?
- *Data model*. What is the underlying data model for the ontologies in the tools?
- *Reasoning*. Do the tools verify newly added data and check for consistency when the ontology changes?
- *Example ontologies*. Are example ontologies available? Are they helpful in understanding the tools?
- *Reuse*. Can previously created ontologies be reused?
- *Formats*. Which data formats are compatible with the tool? What formats can be imported and exported?

- *Visualization.* Do the users get a good overview over the ontology and its components?
- *Help functionality.* Is help functionality available? Is it easy to use?
- *Shortcuts.* Are shortcuts for expert users provided?
- *Stability.* Did the tool crash during the evaluation period?
- *Customization.* Can the user customize the tool and in what way?
- *Extendibility.* Is it possible to extend the tools?
- *Multiple users.* Can several users work with the same tool at the same time?
- *User interface.* The aspects studied in the evaluation are relevance, efficiency, attitude and learnability. Relevance measures how well a user's needs are satisfied by the tool. Efficiency measures how fast users can perform their tasks using the tool. The subjective feelings towards the tools are measured by attitude. Finally, learnability measures how easy or difficult it is to learn the tool for initial use as well as how well a user can remember how to work with the tool.

According to this evaluation, no system is preferred in all situations. All systems have their strengths and weaknesses. The main strengths of Protégé-2000 compared to the other systems are its user interface, the extendibility using plug-ins, the functionality that the plug-ins provide (such as merging) as well as the different formats that can be imported and exported. Chimaera's main strengths are its functionality, including merging and diagnosis, the different formats that can be imported and exported, its help functionality, the shortcuts for expert users and the fact that multiple users can work with the same ontology. Its user interface is its main weakness. The main advantage of OilEd is the fact that its model is description logic-based and that the underlying FaCT system can perform reasoning tasks such as classification and consistency checking. DAG-Edit was specifically built for GO ontologies and has an interface that is easy to use and learn.

Protégé-2000 with PROMPT and Chimaera were also evaluated as ontology merging tools [26]. The test cases used GO ontologies and Signal-Ontology [47]. Both systems create suggestions for operations, such as the merging of two concepts, and compute the additional changes that follow from these operations. Chimaera also suggests taxonomy areas that are candidates for reorganization. The user interface of Protégé-2000 with PROMPT was considered better than Chimaera. It gave a better overview over the ontologies and it was easier to work with. Chimaera, however, provides more functionality and better help facilities. It is also much faster to merge ontologies with Chimaera. This may be an important factor when merging large ontologies. The quality of the suggestions was measured in terms of precision and recall. Precision measures how many of the suggestions are relevant while recall measures how many of the relevant suggestions the system actually proposed. PROMPT had high precision but low recall for the test cases. Chimaera had slightly better recall but low precision. Extensions to the basic algorithm that may improve these results are under development. One such extension is Anchor-PROMPT [37] where similarities in the structure of the ontologies are used to generate new suggestions. Based on the lessons learned in this evaluation the SAMBO system was proposed. SAMBO obtained better results regarding precision and recall than PROMPT and Chimaera for the same GO and Signal-Ontology test ontologies [29].

In [52] Protégé-2000 was assessed as a tool for maintaining and developing the GO ontologies. First, GO was translated into Protégé-2000. Then, the knowledge base management functionality of the system was applied to GO. The results of the assessment suggested that this functionality was useful for checking ontological consistency. The PROMPT tool was used for tracking changes in GO. It found, for instance, concepts that changed from one release to another but had a very similar structure (e.g., they had the same subclasses and superclasses). In this case, PROMPT suggested that these concepts were related. Further, Protégé-2000 was used to make changes to and extend GO.

Some evaluations of ontology tools have been performed in other areas. In [14] Ontolingua, WebOnto, ProtégéWin, OntoSaurus, ODE and KADS22 were compared. The authors used ontologies concerning academia and university studies for testing. It was concluded that for less experienced users WebOnto and ProtégéWin are better suited, while Ontolingua and OntoSaurus might give better support for creating complex ontologies. In [13] requirements were defined for industrial strength ontology management. Scalability, reliability, security, internationalization and versioning were considered to be the most important requirements. Scalability and reliability are needed to support the distributed and cooperative development of large-scale ontologies by multiple users. Security management is needed to protect data integrity, to prevent unauthorized access and to provide access control mechanisms. Internationalization refers to the fact that users at different geographical locations may need to access the same ontologies and this possibly in different natural languages. Finally, versioning is needed as ontologies change and evolve over time. Chimaera, Protégé, WebOnto and OntoSaurus were evaluated with respect to these criteria. No single tool met the requirements. Some of the tools did meet or surpass other requirements such as ease of use, knowledge representation and merging, but these were not considered to be the most important for industrial strength ontology tools. In [38] evaluation criteria for mapping or merging tools were proposed. First, an evaluation should be driven by pragmatic considerations: input requirements, level of user interaction, type of output and content of output. Tools that satisfy a user's pragmatic requirements can then be compared with respect to a performance criterion based on precision and recall. Protégé-2000 with PROMPT was evaluated according to these criteria. A larger survey of ontology tools and methodologies was performed by the OntoWeb Consortium [9, 10].

Regarding ontology development tools the following evaluation criteria were proposed: interoperability with other tools, knowledge representation, inference services and usability. A number of tools were compared according to these criteria. Most of the new tools were found to export and import different markup languages, but there does not exist a comparative study about the quality of the translators. With respect to knowledge representation two families were found: description logics based tools and tools that are based on frames. Most tools provided some inferencing, but only the description logics based tools performed automatic classifications. Finally, regarding usability WebOnto had the most advanced features related to cooperative construction of ontologies. Test experiments are planned for OilEd, OntoEdit, Protégé-2000, WebODE and WebOnto.

7 Conclusion

In this chapter we presented different currently used definitions of ontology. We briefly introduced ontologies in bioinformatics and systems biology and gave examples of how they are currently used. The focus in bioinformatics and systems biology has been on the construction of a community reference as well as on creating a vocabulary for annotation. Ontologies will still be used in this way in the future, while the other uses such as ontology-based search and ontology-based integration of information sources will gain in importance. Regarding representation the focus has been on controlled vocabularies. However, to be able to use the ontologies in more advanced ways, more expressive representation formalisms with reasoning capabilities will need to be used. We also discussed some ontology engineering tools. As was shown in several evaluations, the current ontology engineering tools are a good start, but research is still needed to be able to develop high-quality industrial-strength engineering tools.

Main abbreviations and acronyms

DAG: Directed Acyclic Graph.
DAML: DARPA Agent Markup Language. Developed as an extension to XML and RDF.
DAML+OIL: Semantic markup language for web resources. Based on DAML-ONT and OIL.
DAML-ONT: DAML Ontology language.
EcoCyc: Encyclopedia of Escherichia coli K12 Genes and Metabolism. Scientific database for the bacterium Escherichia coli K12 MG1655.
F-Logic: Frame Logic. Language dealing with most of the structural aspects of object-oriented and frame-based languages.
GO: Gene Ontology consortium. The goal of the Gene Ontology consortium is to produce a controlled vocabulary that can be applied to all organisms.
MeSH: Medical Subject Headings.
MGED: Microarray Gene Expression Data. The MGED Society is an international organization of biologists, computer scientists, and data analysts that aims to facilitate the sharing of microarray data generated by functional genomics and proteomics experiments.
OBO: Open Biological Ontologies. Umbrella web address for ontologies for shared use across different biological domains.
OIL: Ontology Inference Layer. Web-based representation and inference layer for ontologies, which combines the widely used modeling primitives from frame-based languages with the formal semantics and reasoning services provided by description

logics.
OilEd: Ontology editor for OIL.
OKBC: Open Knowledge Base Connectivity. Application programming interface for frame-based knowledge representation systems.
OWL: Web Ontology Language.
PROSITE: Database of protein families and domains.
PROMPT: Plug-in for the Protégé-2000 system supporting ontology merging.
Protégé-2000: Ontology development system.
RDF: Resource Description Framework. The RDF specifications provide a lightweight ontology system to support the exchange of knowledge on the Web.
SAMBO: System for Aligning and Merging Bio-Ontologies.
SBML: Systems Biology Markup Language. An XML-based format for representing biochemical reaction networks.
SHIQ(D): An expressive description logic.
SHOQ(D): An expressive description logic.
SOFG: Standards and Ontologies for Functional Genomics. SOFG is both a meeting and a website. It aims to bring together biologists, bioinformaticians, and computer scientists who are developing and using standards and ontologies with an emphasis on describing high-throughput functional genomics experiments.
STAR/mmCIF: An ontology for macromolecular structure.
Swiss-Prot: A curated protein sequence database.
TAMBIS: Transparent Access to Multiple Bioinformatics Information Sources.
UML: Unified Modeling Language.
WordNet: A lexical database for the English language.
XML: Extensible Markup Language. Web language.

Acknowledgments

We thank Nahid Shahmehri, He Tan, Vaida Jakonienė, Bo Servenius and Jan Maluszynski for discussions and comments. We also acknowledge the financial support of the Center for Industrial Information Technology (CENIIT) and the EU Network of Excellence REWERSE (Sixth Framework Programme project 506779).

References

1. F. Al-Shahrour, R. Diaz-Uriarte, and J. Dopazo. Fatigo: a web tool for finding significant associations of gene ontology terms with groups of genes. *Bioinformatics*, 20(4):578–580, 2004.
2. F. Baader, D. Calvanese, D. McGuinness, D. Nardi, and P. Patel-Schneider. *The Description Logic Handbook*. Cambridge University Press, 2003.
3. P. Baker, C. Goble, S. Bechhofer, N. Paton, R. Stevens, and A. Brass. An ontology for bioinformatics applications. *Bioinformatics*, 15(6):510–520, 1999.

4. S. Bechhofer, I. Horrocks, C. Goble, and R. Stevens. Oiled: a reason-able ontology editor for the semantic web. In *Proc. Joint German Austrian Conference on Artificial Intelligence*, pages 396–408, 2001.
5. V. Chaudhri, A. Farquhar, R. Fikes, P. Karp, and J. Rice. Okbc: A programmatic foundation for knowledge base interoperability. In *Proc. 5th National Conference on Artificial Intelligence*, pages 600–607, Madison, WI, USA, 1998.
6. Chimaera. http://www.ksl.stanford.edu/software/chimaera.
7. F. Collins, E. Green, A. Guttmacher, and M. Guyer. A vision for the future of genomics research. *Nature*, 422:835–847, 2003.
8. Gene Ontology Consortium. Gene ontology: tool for the unification of biology. *Nature Genetics*, 25:25–29, 2000.
9. OntoWeb Consortium. Deliverable 1.3: A survey on ontology tools. 2002. http://www.ontoweb.org.
10. OntoWeb Consortium. Deliverable 1.4: A survey on methodologies for developing, maintaining, evaluating and reengineering ontologies. 2002. http://www.ontoweb.org.
11. DAG-Edit. http://www.geneontology.org/doc/dagedit_userguide/dagedit.html.
12. DAML+OIL. http://www.w3.org/TR/daml+oil-reference.
13. A. Das, W. Wu, and D. McGuinness. Industrial strength ontology management. In *Proc. 1st Semantic Semantic Web Working Symposium*, Stanford, USA, 2001.
14. A.J. Duineveld, R. Stoter, M.R. Weiden, B. Kenepa, and V.R. Benjamins. Wondertools? a comparative study of ontological engineering tools. In *Proc. 12th Workshop on Knowledge Acquisition, Modeling and Management*, Banff, Canada, 1999.
15. EcoCyc. http://www.ecocyc.org/.
16. FaCT. http://www.cs.man.ac.uk/∼horrocks/FaCT.
17. GO. http://www.geneontology.org.
18. A. Gómez-Pérez. Ontological engineering: A state of the art. *Expert Update*, 2(3):33–43, 1999.
19. T. Gruber. A translation approach to portable ontology specification. *Knowledge Acquisition*, 5(2):199–220, 1993.
20. N. Guarino and P. Giaretta. Ontologies and knowledge bases: Towards a terminological clarification. In N. Mars, editor, *Towards Very Large Knowledge Bases: Knowledge Building and Knowledge Sharing*, pages 25–32. IOS Press, 1995.
21. I. Horrocks. Fact and ifact. In *Proc. International Workshop on Description Logics*, pages 133–135, Linköping, Sweden, 1999.
22. I. Horrocks, D. McGuinness, and C. Welty. Digital libraries and web-based information systems. In F. Baader, D. Calvanese, D. McGuinness, D. Nardi, and P. Patel-Schneider, editors, *The Description Logic Handbook*, pages 427–449. Cambridge University Press, 2003.
23. M. Hucka, A. Finney, H.M. Sauro, H. Bolouri, and J. Doyle et al. The systems biology markup language (sbml): a medium for representation and exchange of biochemical network models. *Bioinformatics*, 19(4):524–531, 2003.
24. R. Jasper and M. Uschold. A framework for understanding and classifying ontology applications. In *Proc. 12th Workshop on Knowledge Acquisition, Modeling and Management*, Banff, Canada, 1999.
25. M. Kifer, G. Lausen, and J. Wu. Logical foundations of object oriented frame based languages. *Journal of the ACM*, 42:741–843, 1995.
26. P. Lambrix and A. Edberg. Evaluation of ontology merging tools in bioinformatics. In *Proc. Pacific Symposium on Biocomputing*, pages 589–600, Kauai, Hawaii, USA, 2003.
27. P. Lambrix, M. Habbouche, and M. Pérez. Evaluation of ontology development tools for bioinformatics. *Bioinformatics*, 19(12):1564–1571, 2003.

28. P. Lambrix and V. Jakonienė. Towards transparent access to multiple biological databanks. In *Proc. 1st Asia-Pacific Bioinformatics Conference*, pages 53–60, Adelaide, Australia, 2003.
29. P. Lambrix and H. Tan. Merging daml+oil ontologies. In *Proc. 6th International Baltic Conference on Databases and Information Systems*, Riga, Latvia, 2004.
30. D. McGuinness, R. Fikes, J. Rice, and S. Wilder. The chimaera ontology environment. In *Proc. 17th National Conference on Artificial Intelligence*, pages 1123–1124, Austin, TX, USA, 2000.
31. D. McGuinness, R. Fikes, J. Rice, and S. Wilder. An environment for merging and testing large ontologies. In *Proc. 7th International Conference on Principles of Knowledge Representation and Reasoning*, pages 483–493, Breckenridge, CO, USA, 2000.
32. MeSH. http://www.nlm.nih.gov/mesh/meshhome.html.
33. MGED. http://mged.sourceforge.net/.
34. R. Neches, R. Fikes, T. Finin, T. Gruber, T. Senator, and W. Swartout. Enabling technology for knowledge engineering. *AI Magazine*, 12(3):26–56, 1991.
35. N.F. Noy, R. Fergerson, and M. Musen. The knowledge model of protégé-2000: Combining interoperability and flexibility. In *Proc. 12th International Conference on Knowledge Engineering and Knowledge Management*, pages 97–112, Juan-les-pins, France, 2000.
36. N.F. Noy and M. Musen. Prompt: Algorithm and tool for automated ontology merging and alignment. In *Proc. 17th National Conference on Artificial Intelligence*, pages 450–455, Austin, Texas, USA, 2000.
37. N.F. Noy and M. Musen. Anchor-prompt: Using non-local context for semantic matching. In *Proc. IJCAI01 Workshop on Ontologies and Information Sharing*, 2001.
38. N.F. Noy and M. Musen. Evaluating ontology-mapping tools: Requirements and experience. In *Proc. EKAW Workshop on Evaluation of Ontology Tools*, Siguenza, Spain, 2002.
39. N.F. Noy, M. Sintek, S. Decker, M. Crubézy, R. Fergerson, and M. Musen. Creating semantic web contents with protégé-2000. *IEEE Intelligent Systems*, March/April:60–71, 2001.
40. OBO. http://obo.sourceforge.net.
41. OilEd. http://oiled.man.ac.uk.
42. OWL. http://www.w3.org/2001/sw/WebOnt/.
43. PROMPT. http://protege.stanford.edu/plugins/prompt/prompt.html.
44. Protégé. http://protege.stanford.edu/index.html.
45. R. Stevens R, I. Horrocks, C. Goble, and S. Bechhofer. Building a reason-able bioinformatics ontology using oil. In *Proc. IJCAI Workshop on Ontologies and Information Sharing*, 2001.
46. SBML. http://sbml.org.
47. Signal-Ontology. http://ontology.ims.u-tokyo.ac.jp/signalontology/.
48. SOFG. http://www.sofg.org/resources/key.html.
49. R. Stevens, C. Goble, and S. Bechhofer. Ontology-based knowledge representation for bioinformatics. *Briefings in Bioinformatics*, 1(4):398–414, 2000.
50. J. Westbrook and Ph. Bourne. Star/mmcif: An ontology for macromolecular structure. *Bioinformatics*, 16(2):159–168, 2000.
51. WordNet. http://www.cogsci.princeton.edu/∼wn/.
52. I. Yeh, P. Karp, N.F. Noy, and R. Altman. Knowledge acquisition, consistency checking and concurrency control for gene ontology (go). *Bioinformatics*, 19(12):241–248, 2003.

Natural Language Processing and Systems Biology

K. Bretonnel Cohen and Lawrence Hunter

Center for Computational Pharmacology, University of Colorado School of Medicine, Denver, USA.
E-mail: {kevin.cohen, larry.hunter}@uchsc.edu

Summary. This chapter outlines the basic families of applications of natural language processing techniques to questions of interest to systems biologists and describes publicly available resources for such applications.

1 Introduction

Natural language processing (NLP) is the *processing*, or treatment by computer, of *natural language*, i.e., human languages, as opposed to programming languages. The two differ from each other in a very fundamental way: the interpretation of a programming language is designed not to be ambiguous, while the possible interpretations of natural language are potentially ambiguous at every level of analysis. The processing of computer languages is a subject for computer science and is generally treated in courses on compiler design. In contrast, the processing of natural language crosses a number of disciplines, including linguistics, computer science, and engineering.

One of the more surprising developments in bioinformatics and systems biology has been the attention that NLP has received at bioinformatics conferences in recent years. The Pacific Symposium on Biocomputing (PSB) and Intelligent Systems for Molecular Biology (ISMB) conferences began publishing papers on the topic in the early 1990s devoting entire sessions to the topic in the late 1990s. The natural language processing community has reciprocated, with the Association for Computational Linguistics offering workshops on NLP in the molecular and systems biology domain for the past three years ([56, 2, 48]). This is a welcome turn of events for the NLP community; although the medical domain has long been a target of interest for computational linguists, the medical community has yet to adopt natural language processing in a widespread way. In contrast, the current state of events, in which linguists find biologists coming to them, is a happy one. The results have been beneficial to both groups, with biologists gaining curation tools and linguists taking advantage of the large, well-curated resources that the biological community has made available in recent years. Biologists are increasingly faced with a body of

W. Dubitzky and F. Azuaje (eds.), Artificial Intelligence Methods and Tools for Systems Biology, 147–173.

literature that is too large and grows too rapidly to be reviewed by single researchers. At the same time, it becomes increasingly clear that relevant data is being published in communities outside of the traditional molecular biology subfields. Faced with the need to perform systematic surveys of all published information about multiple genes and proteins returned in large numbers by high-throughput assays, there is a growing awareness among molecular biologists that automated exploitation of the literature may be not just useful, but essential[1].

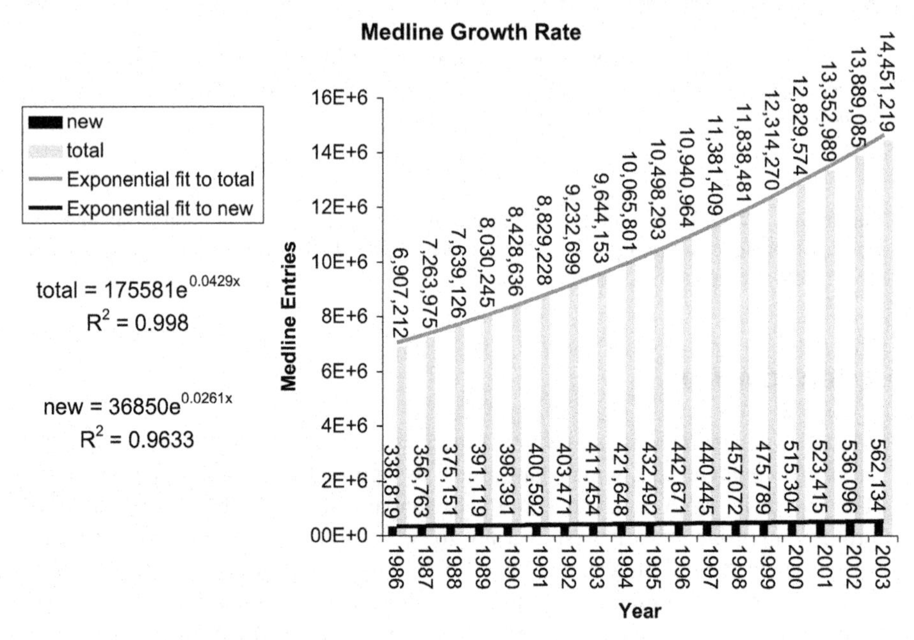

Fig. 1. Growth in Medline over the past 17 years. The hollow portion of the bar is cumulative size up to the preceding year; the solid portion is new additions in that year.

Unfortunately for impatient consumers—perhaps fortunately for curious scientists—NLP is approximately as difficult as it is important. It requires enormous amounts of knowledge, on a number of levels. For example, knowledge of how words are formed (*morphology*) is required to understand words like *deubiquitinization* that are complex and may not have been seen before. Knowledge of how phrases combine (*syntax*) is needed to understand why a sentence like *These findings suggest that FAK*

[1] NLP techniques have also proven useful for macromolecular sequence analysis. This includes the use of hidden Markov models (see e.g., [57] for a general overview and [5] for biological applications), a technique from speech recognition; and the use of phrase structure grammars (see [97] for an excellent review and [96] for a more detailed exposition). In a related vein, techniques from computer science for efficient string searches and tree manipulations have been important as well; these are covered very well in [42]. These are beyond the scope of this chapter, and the reader is referred to the cited references for more information on them.

functions in the regulation of cell migration and cell proliferation ([41]) is ambiguous (does FAK play a role in cell proliferation and in the regulation of cell migration, or does it play a role in the regulation of cell proliferation and in the regulation of cell migration?). These problems are difficult enough—despite the fact that since the 1960s most linguists have been working on the English language, there is still no comprehensive analysis of the syntax of English available. However, they pale next to the difficulty of representing the knowledge about the world that we make use of in understanding language. As human users of language, world knowledge is simultaneously so pervasive and so taken for granted in our understanding of language that we are generally unaware of it and may be difficult to convince that it plays a role at all. Consider the sentences *she boarded the plane with two suitcases* and *she boarded the plane with two engines* ([54]). Both sentences are equally syntactically ambiguous, with two possible phrasal structures for each sentence, depending on whether the plane or the woman has the suitcases or the engines. However, humans are unlikely to entertain two analyses of either sentence—rather, one analysis (and a different one in each case) seems obvious and exclusive for each. This phenomenon is based on knowledge that humans have about the kinds of things that people and airplanes are and are not likely to have. Representing this level of knowledge in the breadth and depth that would be required for understanding unrestricted text in general English or any other language has so far remained an elusive goal.

2 NLP and AI

NLP has a long history in artificial intelligence, and vice versa. There are two main lineages in NLP, one of which traces itself back to Roger Schank (see e.g., [93]). Historically, approaches to conceptual analysis work in AI have tended to be heavily based on semantic processing and knowledge-based techniques, with syntax often (although certainly not always) having a marginal position and sometimes eschewed completely, at least in principle. [103] provides a comprehensible overview of NLP research in the Schankian paradigm through the early 1980s. Since that time, two main trends of thought on NLP have emerged from the AI community: the *direct memory access* paradigm and implementations of conceptual dependency parsing. The *direct memory access* paradigm ([87, 69, 31]) has investigated the implications of thinking of language processing as change in mental representations. This offers some real advantages with respect to mapping the output of NLP to known concepts and entities in an ontology or knowledge base that is lacking in term-based approaches to NLP. (For example, for the input ... *Hunk expression is restricted to subsets of cells...* ([38]), a good term-based system will output the fact that *Hunk* is expressed; a concept-based system might output the fact that LocusLink entry 26 559 is expressed.) It has had some commercial application in the area of robot control by NASA, and shows promise for language processing in the systems biology domain, including in the areas of word sense disambiguation and resolution of syntactic ambiguity ([53]). The conceptual dependency parser ([61, 88]) has had success in the kinds of information extraction tasks that are of much current interest in the bioinformatics community. (Other AI researchers have increasingly pursued statistical approaches to NLP, e.g., [19, 20]). Systems biology literature shows every indication of being the right target at the right time for AI-inspired approaches. Though the necessity of incorporating knowledge into language processing has long been acknowledged, in the past knowledge-based approaches have been thought to be impractical due to both the high cost of knowledge engineering and the breadth and depth of 'common-sense' knowledge required to parse general English. Within just the recent past, the cost argument has ceased to hold as much weight, as the molecular and systems biology community has released for public use large, carefully curated resources like LocusLink ([67] and the Gene Ontology ([24, 25]). With respect to the depth and breadth of knowledge required, we maintain that it is substantially less for molecular biology literature than for general English: nothing you need to know to understand molecular biology is everyday, common-sense knowledge- –everything that anyone knows about molecular biology came from a textbook, a journal article or an experiment. Thus, the time is ripe for applying AI to NLP in systems biology[2].

3 NLP and systems biology

The importance of NLP for systems biology comes from the high-throughput nature of modern molecular biology assays. The drinking-from-a-firehose nature of the business creates the opportunity for fruitful application of NLP techniques in two ways:

- It makes automated techniques for handling the literature attractive by fueling a rate of publication that is unequaled in the history of science, or indeed of the world.
- At the same time, it makes progress in the field of NLP possible by providing a huge body of data in a restricted domain for training and evaluation of NLP systems.

Specific applications of NLP to biological data or assays include automated literature searches on sets of genes returned by an experiment; annotation of gene lists with Gene Ontology concepts; improvement of homology search; management of literature search results; aids to database curation; and database population. These biological tasks have been approached through a variety of NLP techniques, including information extraction, bibliometrics, and information retrieval. In addition, there are subtasks whose successful accomplishment is key to all of these. These include

[2]It should be noted that the molecular biology domain has long been known to be a good target for natural language processing applications due to its good fit to a *sublanguage* model of language. A *sublanguage* is a genre of language use which deals with a semantically restricted domain and has certain qualities which make it more amenable to machine processing than is unrestriced language. A number of recent papers [33] have discussed the fit of systems biology texts to the sublanguage model.

entity identification (see Section 4.2), tokenization (see page 160), relation extraction (see Section 4.3), indexing (see page 165), and categorization and clustering (see page 166). These subtasks are discussed in detail in the sections that follow.

3.1 Where NLP fits in the analysis pipeline

NLP fits into the bioinformatics data analysis pipeline in two ways, or at two points in the process: at the beginning, by aiding in the analysis of the output of high-throughput assays, thus helping the scientist bring a project from experiment to publication; and at the end, by helping the working researcher exploit the flood of publications that fills Medline at the rate of 1500 abstracts a day. (This split in times of application of the technology does not, however, correspond to any division of natural language processing techniques into different categories; as we will see, a given biological application can be implemented using a variety of different NLP technologies, and a single NLP technique may be utilized for a variety of types of biological applications.) We can also think of NLP techniques as helping the biologist approach two kinds of tasks: on the one hand, ad hoc location of data about single items of interest to them (where *single* might be a single gene or protein, or the output of an experiment, which might itself be a list, for example of differentially expressed genes). In this case, the strength of NLP is its potential for mining information from communities whose literature the researcher might not be aware of but whose work has relevance to her (consider for instance the newly discovered importance of a pregnancy-related protein in heart failure [29]). The other type of task can be thought of as systemic in nature, for example population of databases or aiding database curators; here we make more sweeping queries, of the nature of *tell me about every protein-protein interaction described in the literature.*

3.2 Database population and curation

Rapid population of databases of biologically interesting information was an early motivation for NLP in bioinformatics. The idea that if protein names could be located in text, then we could automatically populate databases of facts about proteins—for example, their interactions with other proteins, as in DIP and BIND—comes up in the first of the modern papers on molecular biology NLP, [35]. The problem to be solved is that enormous amounts of information on the topic are present in the systems biology literature, and we would like to convert that free-text information into a computable form, i.e., entries in structured databases. This is doable manually, but at great cost in terms of time and financial resources. Two basic approaches have been suggested—bibliometric techniques, and information extraction techniques.

Database population belongs to a class of problems in which the goal of NLP is to discover a very limited range of types of facts —perhaps only one. A typical example is protein-protein interactions. Bibliometric approaches are based on the assumption that if two proteins are mentioned in the same text (typically an abstract), then there might be a relationship between them. The PubGene system ([55]) is a good example of such a system. Sophisticated approaches like PubGene attempt to normalize for the fact that two proteins might be mentioned in the same text by chance. They typically find only pairwise interactions, an exception being the AlcoGene module of the INIA web site, which finds interactions of arbitrarily large arity. In general, bibliometric approaches suffer from problems related to entity identification. Either they are restricted with respect to the kinds of entity referents that they find—for example, PubGene utilizes only gene symbols and single-word names—or they are swamped by false positives due to synonymy issues, or both. [116] has a good discussion of sources of false positives in bibliometric approaches.

Information extraction offers a more constrained approach to database population. Examples of papers whose stated goal is database population using information extraction techniques include [8, 26]. Information extraction targets a very restricted set of types of assertions, and hence is less susceptible to the extreme low precision problems of bibliometric systems. In general, no technique has proven sufficiently accurate for completely automated population of databases. However, a number of techniques produce output that is of sufficient quality to aid human curators of such databases. Systems biology databases that store information that is more abstract than sequence data, such as BIND, Swiss-Prot, and OMIM, are typically hand-curated by experienced scientists, with new entries coming from findings reported in the scientific literature. A growing body of work addresses the needs of such curators for a fast and efficient way to navigate or filter the high volume of publications that characterizes the rapid rate of progress in systems biology today. The potential utility of NLP in curation efforts is so apparent that some recent competitions have been funded or materially aided by various databases.

3.3 Aids to analysis of high-throughput assays

Gene expression arrays

A number of studies have specifically addressed issues in the analysis of gene expression array data. An early such system was MedMiner ([107]), which was designed to perform automatic literature searches on large numbers of genes found to be of significance in an expression array study. Such studies often result in large lists of genes which may lead to thousands of articles being returned by a literature search; MedMiner helps the experimenters navigate these large bodies of literature by sorting them according to categories known to be of interest to molecular biologists. This work has been extended to other user communities, including researchers on the molecular biology of substance abuse and cancer researchers. Shatkay et al. [98] describes a method for detecting functional relationships in microarray data. Other approaches to the application of NLP to the interpretation of gene expression arrays have concentrated on using literature to augment the classifications of genes already present using the Gene Ontology ([86]).

3.4 Interaction and pathways

A significant body of work has concentrated on the discovery of networks of interactions and on pathway discovery. The interactions are generally between proteins, although other kinds of 'interactions' or associations have been investigated as well, including:

- Proteins and drugs ([90, 107])
- Proteins and diseases ([26, 100, 101])
- Proteins and subcellular locations ([26, 100, 101])

In general, the linguistic and computational problems are the same, regardless of the exact nature of the interaction.

4 Issues and resources in natural language processing

4.1 Evaluation

Metrics

Most evaluation in NLP is done by calculating values for precision, recall, and often F-measure on the output of a system, evaluated against a gold standard. Gold standard data is, in the best-case scenario, data that is hand-annotated by domain experts. It is often constructed by running some automated system against a set of inputs, and then having it hand-corrected by domain experts. It is preferable to have multiple human experts annotate the data. When this is done, inter-annotator agreement can be calculated, e.g., by calculating the κ statistic. This can be an indicator of the difficulty of the task, e.g., indicating the possible upper limit of system performance. Preparing such gold standard data sets is a pressing issue in the systems biology NLP domain. When gold standard data sets are available, they are listed in the relevant subsections.

Precision measures how often the system is correct when it outputs a particular value. It is similar to specificity, and is calculated by dividing the number of correct outputs (*true positives*, or *TP*) by the total number of outputs. The total number of outputs is the number of correct outputs plus the number of incorrect outputs (*false positives*, or *FP*), so the equation is often given as $P = TP/(TP + FP)$. *Recall* measures how often the system correctly finds the right things to output. It is similar to sensitivity, and is calculated by taking the ratio of correct outputs by the total number of potential correct outputs. The total number of potential correct outputs is the number of correct outputs plus the count of things that should have been output but were not (*false negatives*, or *FN*, so the equation is often given as $R = TP/(TP + FN)$. The *F-measure* or *harmonic mean* attempts to balance the contributions of precision and recall to system performance. It is calculated by $2PR/(P+R)$.[3] [54] provides a cogent overview of these and other metrics for evaluating NLP systems.

Precision and recall are taken from the information retrieval (IR) community. The prototypical IR task is to retrieve from some set of documents all and only the

[3]The F-measure can be calculated in other ways that allow for weighting precision more or less, relative to recall—see [57], and [68] pp. 268–270.

documents that are relevant to some query. We assume that the set includes some documents that are relevant, and some that are not. Documents that are relevant and are successfully retrieved by the system are thus 'true positive' outputs. Documents that are retrieved by the system but that actually are *not* relevant to the query are 'false positive' outputs, and documents that truly are relevant to the query but that the system failed to retrieve are 'false negative' outputs.

Bake-offs

Most systems currently described in the literature were evaluated with locally prepared data, and sometimes with idiosyncratic scoring methods. However, in recent years the systems biology NLP community has experimented with its first 'bake-off'-style competitions, where each participating group is evaluated on the same data, with outputs being scored at a central location using consensus criteria. The recent competitions have been:

- the KDD Cup genomics challenge, described in [119]
- the TREC 2003 genomics track, described in [47]
- BioCreative, described at [6]

4.2 Entity identification

All applications of NLP to problems in computational and systems biology require the ability to recognize references to genes and gene products in text. For example, in the sentence fragment *association of ADHD with DRD4 and DRD5* ([60]), we want to know the DRD4 and DRD5 are genes, but ADHD is not, despite the fact that all three words look very much the same. The general problem of recognizing things of a particular class in free text is known as *entity identification* or *named entity recognition*. The problem was first defined in the general-language domain in the context of the Message Understanding Conferences ([78, 54]). Entity identification has been a topic of interest in the systems biology NLP domain for about as long as NLP has been of interest to systems biologists, and in fact the most heavily cited NLP paper from a computational bioscience conference, [35], was on this topic. In general-language domains, the set of entities has tended to be fairly heterogeneous, ranging from names of individuals to monetary amounts; in the systems biology domain, the set of entities is sometimes restricted to just genes and gene products, with individual authors tending to define the task on an ad hoc basis. (The BioCreative competition may bring about some standardization of the task definition.)

Approaches to entity identification in the systems biology domain fall into two general classes: rule-based approaches, and machine-learning-based approaches (see below). Rule-based approaches generally rely on some combination of regular expressions (see paragraph below) to define patterns that match gene names, and some logic for extending names to the right and/or left. For example, a rule-based approach might use a regular expression such as `/^[a-z]+[0-9]+$/` (any sequence of one or more lower-case letters followed immediately by any sequence of one or more digits) to recognize that p53 is a gene name. It might also include a rule (possibly also implemented as a regular expression) to include the word gene if it occurs immediately to the right of a string that is recognized by that pattern. In addition to Fukuda et al.'s work, examples of rule-based approaches in the literature include [79]. Fukuda's PROPER system is a rule-based system that is freely available for download at [58].

A variety of machine-learning-based approaches to entity identification have been tried. These are mostly the work of the Tsujii lab. Approaches have included decision trees, Bayesian classifiers, hidden Markov models, iterative error reduction, and support vector machines. Tanabe and Wilbur's ABGene system is a learning-based system that is freely available for download at [1].

Almost all work on entity identification can be described as entity 'location.' The task is generally defined as locating entities in text. There is generally no attempt to map the entities that have been located to a database of genes or gene products, despite the benefits to being able to do this. This more complex task may be referred to as concept identification. [21] addresses some of the problematic issues for this task from a structural linguistic perspective. BioCreative task 1B addressed the issue.

Entity identification systems that attempt to rely solely on 'look-up' in a dictionary or gazetteer of names typically perform quite poorly, with coverage generally only in the range of 10-30%, meaning that only 10-30% of the gene names in a corpus can typically be found this way, even allowing for some variability in the form of the names between the reference source and the corpus, such as letter case, hyphenation, etc. (see [21] for a discussion of such variability).

A *regular expression* is a mathematical formula for specifying the class of objects that belong to a particular set. When applied to natural language processing, the objects are textual strings. Regular expression engines typically allow for making reference to specific positions in a string, for allowing choices between a set of characters, for repetition, and for optionality. For example, in the regular expression `/^[a-z]+[0-9]+$/` the carat species the beginning of the string, `[a-z]` represents a choice between any of the lower-case letters, `[0-9]` represents a choice between any of the digits, the plus-signs indicate that the 'choice' that precedes it can be repeated any number of times, and the dollar-sign specifies the end of the string. Taken together, the elements of the regular expression specify strings that begin with one or more letters and end with one or more digits. Thus, the set of strings that is specified by the regular expression includes *p53, pax9,* and *hsp60.* Chapter 2 of [57] gives an excellent introduction, both theoretical and applied, to regular expressions. [50] provides an excellent introduction to the use of regular expressions in the Perl programming language; because most modern regular expression engines mimic Perl's syntax, much of its material is applicable to other languages as well.

Resources for entity identification

Resources for entity identification fall into two classes:

- Lists of names and symbols for 'dictionary' construction

- Software for performing entity identification

At this writing, a variety of publicly available sources for dictionary construction exist. Most provide both names and symbols, and some also provide synonym lists. These include the following:

- LocusLink's LL_tmpl file. LocusLink ([67]) supplies an extremely large number of names and symbols, including synonyms for each, often multiple ones. These are available for a wide variety of species (thirteen at time of writing). There is no attempt at standardization. From the point of increasing recall, this is a benefit. Names and symbols can be extracted from a number of fields, including
  - OFFICIAL_GENE_NAME
  - PREFERRED_GENE_NAME
  - OFFICIAL_SYMBOL
  - PREFERRED_SYMBOL
  - PRODUCT
  - PREFERRED_PRODUCT
  - ALIAS_SYMBOL
  - ALIAS_PROT

  It is available for downloading at [65]. Java classes for parsing the data file and representing LocusLink entries are available from the authors.
- HUGO: The Human Gene Nomenclature Database supplies a much smaller number of names and symbols for human genes. Some symbols are provided. The symbols are standardized. It is described in [114] and is available for downloading in a variety of formats at [51].
- FlyBase provides names, symbols, and synonyms for D. melanogaster genes. [49] and [77] discuss its use in NLP.

Finally, the reader should consult the 'interesting gene name' site at [39]; for comic relief, be sure to note the 'worst gene names' page. See also the FlyNome site (`http://www.flynome.org`) for explanations of some of the more interesting Drosophila names.

Software for performing entity identification falls into two classes—systems that are available over the Internet for remote usage, and systems that the user installs locally. Availability of the former type of system of course varies. At the time of writing, the following over-the-Internet systems are available:

- The GAPSCORE system is available via a Web-based interface at [37]. It returns a list of potential gene names in the input text with a score that rates the probability that each name is a gene. (It is also available through an XML-RPC interface from a variety of languages—see below.) It is described in [18].
- Yapex is available via a Web-based interface at [118]. Yapex has the unusual feature of being able to use information about names mentioned more than once in an input to improve recognition of those names on subsequent mentions. It is described in [32].

- The Descriptron system, under development at the Center for Computational Pharmacology and described in [74], provides (among other services) look-up for gene symbols, names, and other identifiers, allowing rapid determination of whether or not an identifier is ambiguous as to species or as to the underlying sequence data.

The following systems for local installation are available:

- Chang et al.'s GAPSCORE system is available at [36].
- The ABGene system is available for download at [1]. It performs two functions at once: it simultaneously locates named entities, and performs part-of-speech tagging, such that all non-entities in the output have POS tags in place. It is available on Solaris and Linux; installation on Linux requires Slackware (a specific Linux distribution, available at [102]). It is described in [108] and [109].
- The KeX/PROPER system is available for download at [58]. It produces SGML-style output. It is optimized for yeast. It is described in [35].

Evaluation of entity identification

Two kinds of data sets for evaluation of entity identification systems exist. One kind is data sets assembled by individual authors and made available in conjunction with their publications. Recently, another kind of data set has become available, as well—publicly available, carefully-curated large data sets intended for use in challenge tasks. These latter may become standard data sets for publication purposes.

- The GENIA corpus is an extensively hand-annotated corpus of abstracts on human blood cell transcription factors. It is split into sentences and the content is fully tokenized[4]. It is part-of-speech tagged, and is also annotated with respect to a sophisticated ontology of the molecular domain. This ontology includes a number of concepts that correspond to named entities as that term is used in this chapter, i.e., genes and gene products. It is the largest corpus of its type currently available, comprising 2 000 abstracts with 18 545 sentences containing 39 373 named entities. It is available at [40] and is fully described in [81, 59].
- The BioCreative corpus comprises 10 000 sentences and titles with 11 851 named entities. Unlike the GENIA corpus, it was deliberately constructed to be heterogeneous (within the constraints of the molecular biology domain). It includes sentences that contain deliberately challenging false positives. It is downsampled from abstracts, which removes some classes of contextual cues. The corpus was originally constructed by the National Library of Medicine. It was made publicly available in conjunction with the BioCreative comptetition on entity identification. It is available for download at [6].
- the Yapex data set—about 200 Medline abstracts, some of which are a re-annotated subset of the GENIA corpus. It is available for download at [117].

[4]*Tokenization* is the separation of input into appropriately sized chunks for analysis. The term often refers to separating words and punctuation into individual *tokens* (see the example on page 160). *Sentence tokenization* is the separation of input into sentences.

- The authors make available a system for generating test data for entity identification systems at [52]. The system allows the user to generate customized test suites to evaluate performance on different types of names and symbols in a variety of sentential contexts. It is described in [23].

4.3 Information extraction

Information extraction (IE) is the location of assertions about restricted classes of facts in free text. It is also sometimes referred to as relation extraction. IE can be thought of as a 'robust' approach to natural language understanding ([57]) in that rather than trying to build a system that 'understands' all aspects of an input text, workers in information extraction try to 'understand' only assertions of a very restricted sort. For example, an early system in the molecular biology domain extracted assertions about subcellular localization of proteins ([26]). Information extraction technologies have a wide range of applications. The most basic of these uses the results of information extraction directly to populate a knowledge base. Extracted assertions can also be used as input data for other NLP-based applications, such as ontology construction, network discovery, and information retrieval. (So far the immaturity of the technology has stood in the way of success in such efforts.)

Approaches to information extraction can be classified into two broad categories—rule-based, and machine-learning-based. In general, rule-based systems tend to apply some linguistic analysis; in contrast, learning-based systems tend to apply less linguistic analysis and to use simpler representations[5]. The first application of information extraction to the molecular biology domain was a rule-based system for finding protein-protein interactions, described in [8]. A representative example of a rule-based system is described in [83]. These authors developed a set of regular expressions defined over part-of-speech (POS) tags and entities that perform some analysis of sentence structure, such as recognizing complex coordinated sentences, and then recognize simple assertions about protein-protein interactions involving a limited number of verbs and deverbal nouns. Commonly used 'keywords' in these systems (see e.g., [8, 10]) include:

- *interact*
- *associate*
- *bind*
- *complex*
- *inhibit*

[5]A very common model for representing a text in a machine learning framework is the *bag of words* (BOW). In a BOW model, the text is represented as a vector in which each element represents a single word. (The value for the element may be binary, i.e., indicating presence or absence of the word, or it may be weighted in some way.) The BOW metaphor takes its name from the fact that the features reflect nothing but the words, crucially excluding order—thus, the BOW representation for the sentences *A upregulates B* and *B upregulates A* would be identical—something like *A:1 B:1 upregulates:1.*

- *activate*
- *regulate*
- *encode*
- *function*
- *phosphorylate*

The first machine-learning-based information extraction system in the molecular biology domain is described in [26]. They developed a Bayesian classifier which, given a sentence containing mentions of two items of interest, returns a probability that the sentence asserts some specific relation between them. For example, given a sentence containing the name of a protein and the name of a cellular compartment, it returns the probability that the sentence asserts that that protein is localized to that cellular compartment. Later systems have applied other technologies, including hidden Markov models and support vector machines.

Things that make information extraction difficult

A variety of factors conspire to make information extraction difficult. These factors fall into two general groups: issues that must be dealt with in most information tasks, and issues that may be specific to the systems biology domain. Entity identification is frequently cited in error analyses as a source of low recall: inability to solve the entity identification problem leads to missed assertions. *Coordination,* the linking of structures by words like *and* and *or,* is another problematic phenomenon. Negation is often simply ignored, a notable exception to this being the work reported in [62] and [63]. *Anaphora,* or references to entities that have been named earlier, often by words like *it,* are another source of low recall.

Low-level linguistic analysis and preprocessing

Many issues of low-level linguistic analysis arise in information extraction. These include:

- sentence tokenization
- word-level tokenization
- entity identification
- part-of-speech tagging
- stemming
- abbreviation expansion

The National Library of Medicine makes available a variety of tools that might be of use. These include:

- *lvg* (lexical variant generation), a set of Java API's for normalizing and generating variant forms of biomedical terms. lvg is described in [28] and is available for download at [66].
- *MetaMap*, a system for finding biomedical concepts in free text. MetaMap is described in [3] and is available for download at [76].

Sentence tokenization

Sentence tokenization is the process of separating a chunk of text into individual sentences. A problem with tokenization of sentences in molecular biology text is that case is not always a reliable indicator of sentence boundaries. Consider for example the following text, which should be split into four sentences, indicated here by line breaks:

> *Misshapen (Msn) has been proposed to shut down Drosophila photoreceptor (R cell) growth cone motility in response to targeting signals linked by the SH2/SH3 adaptor protein Dock.*
> *Here, we show that Bifocal (Bif), a putative cytoskeletal regulator, is a component of the Msn pathway for regulating R cell growth cone targeting.*
> *bif displays strong genetic interaction with msn.*
> *Misshapen (Msn) has been proposed to shut down Drosophila photoreceptor (R cell) growth cone motility in response to targeting signals linked by the SH2/SH3 adaptor protein Dock.*

The final sentence of the selection (from the abstract of [91]) begins with a mention of the recessive form of the Bifocal gene. The authors have followed the Drosophila community's convention of indicating dominance/recessiveness of an allele by using upper case for the initial letter of the name/symbol when discussing the dominant form, and lower case for the recessive allele. Other difficulties come from domain-specific entities that can contain internal punctuation that would normally be sentence-final, such as chromosomal locations (*p24.2*), species names (*S. cerevisiae*), etc.

Approaches to sentence tokenization can be divided into two categories: rule-based, and learning-based. Appendix B of [16] gives a set of heuristics for rule-based sentence tokenization of Medline abstracts. No publicly distributed tools that are customized for the molecular biology domain are currently available.

Word-level tokenization

Most NLP projects require breaking the input into word-sized chunks. The definition of what counts as a *word* is often frustratingly domain-specific. Molecular biology text provides its own challenges in this regard. For example, tokenization routines generally split punctuation from the words to which it is attached. They generally count hyphens as separable punctuation. However, this often yields undesirable results on molecular biology text. Consider, for example, the sentence *Relaxin, a pregnancy hormone, is a functional endothelin-1 antagonist: attenuation of endothelin-1-mediated vasoconstriction by stimulation of endothelin thp-B receptor expression via ERK-1/2 and nuclear factor-kappaB* [29]. The desired output of tokenization is shown in Table 1, where it is contrasted with the output of a typical tokenizer. A number of problems with the typical output are apparent. *Endothelin-1-mediated* should be separated into *endothelin-1* and *mediated*, but *endothelin-1* should be kept as a single 'token.' Similarly, *thp-B* is split into three tokens, when it should be maintained as a single unit, and *ERK-1/2* is split into five units. Some tokenization routines actually discard punctuation, including hyphens; this is problematic in the biomedical domain, where e.g., hyphens can be used to indicate negation ([105]) and electrical charge.

Part-of-speech tagging

Part-of-speech tagging is the assignment of part-of-speech labels to individual tokens in a text. The set of labels is typically much larger than the eight categories (noun, verb, preposition, etc.) typically taught in traditional grammar. A common *tagset* (set of tags) includes around forty categories, and much larger sets are known, as well. ([57]:Appendix C gives several pages of tags.) The increased size of NLP tagsets as compared to the eight traditional parts of speech comes in part from finer granularity—for example, where traditional grammar has the category *noun*, a commonly used NLP tagset has the categories *NN (singular or mass noun), NNS (plural noun), NNP (singular proper noun),* and *NNPS (plural proper noun).* It is a challenging task because even within a homogeneous domain, a word can have multiple parts of speech. For instance, in the molecular biology domain, *white* can be an adjective, as in ... *Morgan's awarenesss that white eye-color was not the only genetically determined alternative to red eye-color...* ([30]); a mass noun, as in ... *the appearance of a traite, such as color, was due to the presence of a gene, and white, i.e., no color, to its absence* (op cit); and of course a proper noun. Information extraction systems generally apply a POS tagger and entity identification system as their first steps, in one order or the other. Publicly available entity identification systems are discussed above in 4.2. Publicly available POS taggers include the following:

- Brill: the Brill part-of-speech tagger ([14]) is possibly the most widely used piece of NLP software in the world. It is shipped with data for tagging general English, but can be trained on molecular biology data and has been widely applied to such. It is available at [13].
- TnT: The *Trigrams'n'Tags* part-of-speech tagger ([12]), also known as *TnT*, is a very fast and stable part-of-speech tagger that is available on a variety of platforms. It has been tested on multiple languages, and has an intuitive interface. It is available at [111].

Stemming

It is often useful to be able to determine the stem of words in the input text. A word's *stem* is the main part of the word, exclusive of parts that are added to it to mark plurality, tense, etc. For example, *interact* is the stem of the words *interacts, interacted, interacting,* and *interaction.* Publicly available software for this includes many implementations in a wide variety of languages of the Porter stemmer ([85]), available at [84]. No stemmer has been optimized for NLP in the systems biology domain.

Table 1. Desired and typical outputs of tokenization. The table shows one token per line. Note that the typical tokenization routine tends to break apart things that should remain single units.

DESIRED OUTPUT OF TOKENIZATION	OUTPUT OF A TYPICAL TOKENIZATION ROUTINE
Relaxin	Relaxin
,	,
a	a
pregnancy	pregnancy
hormone	hormone
,	,
is	is
a	a
functional	functional
endothelin-1	endothelin
	-
	1
antagonist	antagonist
:	:
attenuation	attenuation
of	of
endothelin-1	endothelin
	-
	1
-	-
mediated	mediated
vasoconstriction	vasoconstriction
by	by
stimulation	stimulation
of	of
thp-B	thp
	-
	-
	B
receptor	receptor
expression	expression
via	via
ERK-1/2	ERK
	-
	1
	/
	2
and	and
nuclear	nuclear
factor-kappaB	factor
	-
	kappaB
.	.

Lexical resources

Lexical resources are often useful in information extraction, and happily, a number of them are publicly available. (One could argue that the number and high quality of lexical resources that has become available in the recent past make molecular biology the first domain in which knowledge-based NLP has ever been practical.) The advantages of these resources include the ability to recognize multi-word terms, which reduces the amount of low-level parsing necessary ([115]).

- Gene Ontology: the Gene Ontology ([24, 25]) is an ontology of concepts relevant to the systems biology domain. [73, 113, 80] discuss various linguistic aspects of the ontology and its applicability to NLP tasks.
- UMLS: the Unified Medical Language System is a large metathesaurus of biomedical vocabularies. It is documented in [64] and Bodenreider (2004). and [11]. It is available through the National Library of Medicine at [112]. Numerous researchers have investigated its use in natural language processing, including [70, 71, 89, 4, 120, 15, 46, 3, 72], to name just a few.

Abbreviation expansion

The ability to deal with abbreviations is often important in systems biology text. Abbreviations are often defined ad hoc, limiting the usefulness of dictionary-based systems. Additionally, systems biology text also often contains gene symbols. These symbols are often defined in the text in a structure similar to that of an abbreviation definition.

- The BioText project makes Java code for a rule-based system available at [94]. It is straightforward to implement and use, and the authors and others have applied it to the BioCreative Task 1A challenge task. The algorithm is described in [95].
- Chang et al. make a statistically-based system available through a web site and via an XML/RPC server. It can be found at [104]. This system returns a list of potential abbreviation/definition pairs, each with both a categorical and a probabilistic assessment of the likelihood that it is a valid pair. The system is described in [17].

Evaluation of information extraction

There has not yet been a MUC-like competition for information extraction in the molecular biology domain, and so no data set like BioCreative exists yet. Small, generally 'lightly annotated' data sets have been made available by individual researchers. These include:

- Contact Mark Craven at `craven@biostat.wisc.edu` for access to a large dataset of assertions about protein-protein interactions, protein-disease associations, and subcellular localization of proteins.

- Contact Christian Blaschke at blaschke@cnb.uam.es for access to a dataset on protein-protein interactions.

Evaluations of information extraction in this domain typically involve precision, recall, and F-measure, but may differ with respect to the domain over which they are measured. Some authors calculate them on the basis of mentions in text, while other authors calculate them on the basis of the underlying concepts. For example, if a test set contains three assertions to the effect that p27 interacts with CK2, then if we are calculating recall on the basis of mentions, then there are three potential true positives, and any that we miss will count as false negatives. On the other hand, if we are calculating recall on the basis of the underlying concepts, then as long as we find at least one assertion that p27 interacts with CK2, we have no false negatives.

The issue of input size

Most NLP work in systems biology takes abstracts (with their titles) as the basic unit of input, rather than full-text articles. One reason for this is purely practical—until recently, access to full-text articles in easily processable formats has been quite limited. (The PubMed Central collection is one current attempt to make full-text articles available.) A small number of researchers has in fact reported success in working with full-text articles. The GENIES system ([34]) was evaluated on a full-length article, and the winning team in the 2002 Genomics KDD cup [119] used full-text articles to great advantage. [109] discusses some of the difficulties that arise when working with full-length articles rather than abstracts, and [27] evaluates the use of inputs of various sizes in an information extraction task.

Resources: raw data for information extraction

Resources: almost all research in systems biology NLP begins with a query to the Entrez interface to the Pubmed document collection, typically through the well-known Entrez interface. Kevin Rosenberg makes LISP code for accessing PubMed available through the BioLisp organization (http://www.biolisp.org), and the National Library of Medicine makes an API available. These queries themselves fall into the category of information retrieval, the subject of the next section.

A local copy of medline allows for heavier usage and faster access than does the National Library of Medicine interface or API's. The BioText project at the University of California at Berkeley and Stanford makes available Java and Perl code for parsing the data files provided by the National Library of Medicine into a relational database, as well as the associated schemas. The code and schemas are available at [7] and are described in [82].

4.4 Information retrieval

Information retrieval consists of finding subsets of documents in a larger set that are relevant to some query. Originally a problem in library science, it has largely been reconceived as a WWW query task. In the systems biology domain, all of the standard IR problems present themselves. In addition, there is a twist that is peculiar to this domain. A typical Web query may return thousands of documents, of which only a small number are actually relevant to the user—the 'needle in a haystack' problem. In contrast, a typical query about a gene to the Pubmed search engine may return thousands of documents, of which most are relevant to the user. So, the problem in IR for gene expression array studies is not the Google-task of finding a needle in a haystack—the problem is that the whole haystack is made of needles. The issue then becomes: how to organize this mass of documents in such a way as to make it navigable by the user? Relevant research issues include indexing, query construction, clustering/categorization, and visualization, which are discussed in the following sections.

Indexing

Indexing a document collection is the process of determining the set of terms or words within each individual document that should be used when matching that document to a query. Not all words are equally useful for purposes of indexing. For example, *function words* (words that indicate grammatical information only) such as *a, the,* and textitall are usually considered not to be useful for indexing. Since all documents contain them, they are not useful for determining whether or not a particular document should be returned in response to a particular query. In contrast, *content words* (words that express specific semantic concepts), such as *phosphorylate, protein,* and *BMP-4*, are generally good candidates for indexing. Standard mathematical procedures for determing the usefulness of particular words for indexing particular document collections exist—see Salton 1989 and Jackson and Moulinier 2002). To understand why a particular word might be useful for indexing one document collection but not another, consider the word *protein* and two separate document collections: a set of documents about nutrition, and a set of documents about Bone Morphogenetic Protein 4. For the set of documents about nutrition, it is easy to imagine realistic queries for which the presence or absence of the word *protein* in a document will be very useful for deciding whether or not to include that document in the set of documents that are returned. In contrast, for the set of documents that are all about Bone Morphogenetic Protein 4, the presence or absence of the word *protein* is not likely to ever help us decide whether or not to return a particular document.

A number of factors conspire to make indexing for systems biology difficult. These include:

- Massive synonymy of the items of interest. Many of the concepts of interest in systems biology are genes and proteins, which have on average about five synonyms each.
- Multi-word units: Traditional indexing assumes that the unit of interest is a single word. However, the concepts of interest to systems biologists are frequently referenced by multi-word units—review of two corpora of molecular biology texts revealed that about 50% of the mentions of genes and proteins in each corpus were two or more words in length ([22]).

- Difficulty of mapping to a standard ontology: Concepts of interest in systems biology are not limited to genes and proteins, but rather include also concepts such as those described in the Medical Subject Headings (MeSH) and the Gene Ontology. Such ontologies have proven to be useful indexing schemes, but assigning the correct MeSH headings or GO codes is difficult to do automatically, due to synonymy and to variability in the possible forms of multi-word ontology elements. (For example, *regulation of cellular proliferation* can also appear in text as *cell proliferation regulation.*)

  All of these are open research issues.

Clustering and categorization

Clustering and categorization address the issue of taking a set of documents that have been returned in response to a query, and organizing them in a way that helps the user navigate and make sense of them. There are two approaches to clustering and categorization: top-down, and bottom-up.

- *Top-down* clustering organizes a document set according to a pre-existing model of how a user models the conceptual domain. For example, the AlcoGene system is a literature retrieval application for researchers in the molecular biology of alcoholism. It organizes the set of documents returned in response to a query according to which of the following topics they address:
  - nervous system structures
  - behaviors
  - ion channels
  - protein kinases
  - quantitative trait loci

  These categories were arrived at by interviewing domain experts and monitoring their interactions with the literature retrieval system. To see a top-down categorization system in action, try the MedMiner web site, described in [107] and available at [75].
- *Bottom-up* clustering of documents is based on similarities between documents in a set as determined by some metric. Where top-down clustering is based on a priori assumptions about the world to which we map the members of a document set, bottom-up clustering is based entirely on the contents of the documents and requires no model of the world beyond a theory of how to represent document contents and a similarity metric by which to assess them. To see a bottom-up categorization system in action, try Vivisimo's system at `http://www.vivisimo.com`. Limiting the search domain to PubMed, enter the query *p53*. The search returns 200 documents, which Vivisimo separates into a number of categories, including the following:
  - breast (13 documents)
  - activation of caspase (12 documents)
  - hepatocellular carcinoma (10 documents)

Querying with the gene symbol *bmp4* returns a very different set of categories, including:

- neural crest (32 documents)
- tooth (17 documents)
- Tgfbeta (16 documents)
- receptors, fibroblast growth factor (12 documents)

The clusters for the oncogene p53 and the developmental gene bmp4 are quite different. This flexibility is a strength of the bottom-up approach. On the other hand, the clusters are not necessarily relevant to the researcher's interests; the guarantee of relevance is a strength of the top-down approach. [98] presents another perspective on clustering for information retrieval, assuming a usage scenario involving large lists of genes as for example the output of an expression array experiment. Clusters based on papers that are prototypical for particular genes are used to discover functional relationships between genes in a large dataset.

Visualization

Visualization: A very open area of research is visualization of the contents of large document collections. The hypothesis is that users might better be able to navigate large document sets if they have some visual metaphor for the organization of the set, rather than just the flat (or at best hierarchical) lists returned by most literature search interfaces. A good starting point for research in this area is [44]. For a demonstration of an interesting visualization system, see the Pacific Northwest National Lab's ThemeRiver, documented in [43] and viewable at [110].

5 Further reading

In this section I differentiate between general natural language processing, i.e., coverage of the topic that is not specific to a particular genre or domain, and NLP for systems biology. Any investigation of general NLP should start with [57]. For the specific topics of general information extraction, information retrieval, text categorization, entity identification, and summarization, the reader should begin with [54]. [106] describes molecular-biology-specific information retrieval, entity identification, and information extraction systems. For a comprehensive treatment of statistical approaches to general NLP, see [68]. For information retrieval, [92] is a good general text, making up for the fact that it is somewhat dated by the fact that it is incredibly readable; for information retrieval in biomedical domains, [45] is recommended. For NLP in the systems biology domain, some excellent review papers are available, including [9, 116, 99].

Acknowledgments

The authors gratefully acknowledge editing by Todd Kester and help with LaTeX hacking from Sonia Leach.

References

1. ABGene. ftp://ftp.ncbi.nlm.nih.gov/pub/tanabe.
2. S. Ananiadou and J. Tsujii, editors. *Proceedings of the ACL 2003 workshop on natural language processing in biomedicine*. Association for Computational Linguistics, Association for Computational Linguistics, 2003.
3. A.R. Aronson. Effective mapping of biomedical text to the umls metathesaurus the metamap program. In *Proceedings of the AMIA Symposium 2001*, pages 17–21, 2001.
4. A.R. Aronson and T.C. Rindflesch. Query expansion using the umls metathesaurus. In *Proc. AMIA Annu Fall Symp 1997*, pages 485–489, 1997.
5. P. Baldi and B. Søren. *Bioinformatics: the machine learning approach*. MIT Press, 2nd ed edition, 2001.
6. BioCreative. http://www.mitre.org/public/biocreative/2.
7. BioText. http://biotext.berkeley.edu2.
8. C. Blaschke, M.A. Andrade, C. Ouzounis, and A. Valencia. Automatic extraction of biological information from scientific text: protein-protein interactions. In *Intelligent Systems for Molecular Biology 1999*, pages 60–67, 1999.
9. C. Blaschke, L. Hirschman, and A. Valencia. Information extraction in molecular biology. *Briefings in Bioinformatics*, 3(2):154–165, 2002.
10. C. Blaschke and A. Valencia. The frame-based module of the suiseki information extraction system. *IEEE Intelligent Systems*, 17(2):14–20, 2002.
11. O. Bodenreider. Unified medical language system (umls): integrating biomedical terminology. *Nucleic Acids Research*, 32(D):D267–D270, 2004.
12. T. Brants. Tnt—a statistical part-of-speech tagger. In *Proc. of the Sixth Applied Natural Language Processing Conference (ANLP-2000)*, 2000.
13. Brill. Pos tagger site. http://www.cs.jhu.edu/~brill.
14. E. Brill. Transformation-based error-driven learning and natural language processing: A case study in part-of-speech tagging. *Computational Linguistics*, 21(4):543–566, 1995.
15. D.A. Campbell and S.B. Johnson. A technique for semantic classification of unknown words using umls resources. In *Proc AMIA Symp 1999*, pages 716–720, 1999.
16. J.T. Chang. *Using machine learning to extract drug and gene relationships from text*. PhD thesis, Stanford University doctoral dissertation., 2003.
17. J.T. Chang, H. Schütze, and R.B. Altman. Creating an online dictionary of abbreviations from medline. *J Am Med Inform Assoc*, 9(6):612–620, 2002.
18. J.T. Chang, H. Schütze, and R.B. Altman. Gapscore: finding gene and protein names one word at a time. *Bioinformatics*, 20(2):216–225, 2004.
19. E. Charniak. *Statistical language learning*. MIT Press., 1996.
20. E. Charniak. A maximum-entropy-inspired parser. In *Proc. of NAACL-2000*, pages 132–139, 2000.
21. K.B. Cohen, A.E. Dolbey, G.K. Acquaah-Mensah, and L. Hunter. Contrast and variability in gene names. In *of the workshop on biomedical natural language processing*. Association for Computational Linguistics., 2002.
22. K.B. Cohen, P.V. Ogren, S. Kinoshita, and L. Hunter. Entity identification in the molecular biology domain with a stochastic pos tagger. in preparation.
23. K.B. Cohen, L. Tanabe, S.Kinoshita, and L. Hunter. A resource for constructing customized test suites for molecular biology entity identification systems. In *Linking biological literature, ontologies and databases: tools for users*, pages 1–8. Association for Computational Linguistics, 2004.
24. Gene Ontology Consortium. Gene ontology: tool for the unification of biology. *Nature*, 25:25–29, 2000.

25. Gene Ontology Consortium. Creating the gene ontology resource: design and implementation. *Genome Research*, 11:1425–1433, 2001.
26. M. Craven and J. Kumlein. Constructing biological knowledge bases by extracting information from text sources. In *Proc. of the 7th International Conference on Intelligent Systems for Molecular Biology (ISMB-99)*, pages 77–86. AAAI Press, 1999.
27. J. Ding, D. Berleant, D. Nettleton, and E. Wurtele. Mining medline: abstracts, sentences, or phrases? In *Pacific Symposium on Biocomputing 7*, pages 326–337, 2002.
28. G. Divita, A.C. Browne, and T.C. Rindflesch. Evaluating lexical variant generation to improve information retrieval. In *Proc AMIA Symp. 1998*, pages 775–779, 1998.
29. T. Dschietzig, C. Bartsch, C. Richter, M. Laule, G. Baumann, and K. Stangl. Relaxin, a pregnancy hormone, is a functional endothelin-1 antagonist: attenuation of endothelin-1-mediated vasoconstriction by stimulation of endothelin thp-b receptor expression via erk-1/2 and nuclear factor-kappab. *Circ Res*, 92(1):32–40, 2003.
30. R. Falk and S. Schwartz. Morgan's hypothesis of the genetic control of development. *Genetics*, 134:671–674, 1993.
31. W. Fitzgerald. *Building embedded conceptual parsers*. PhD thesis, Northwestern University doctoral dissertation., 1995.
32. K. Franzén, G. Eriksson, F. Olsson, L. Asker, P. Lidén, and J. Cöster. Protein names and how to find them. *International Journal of Medical Informatics*, 67(1-3):49–61, 2002.
33. C. Friedman. Sublanguage—zellig harris memorial. *Journal of Biomedical Informatics*, 35(4):213–277, 2002.
34. C. Friedman, P. Kra, H. Yu, M. Krauthammer, and A. Rzhetsky.
35. K. Fukuda, T. Tsunoda, A. Tamura, and T. Takagi. Toward information extraction: identifying protein names from biological papers. In *Pacific Symposium on Biocomputing 1998*, pages 705–716, 1998.
36. GAPSCORE. Code for gapscore site. `http://bionlp.stanford.edu/webservices.html`.
37. GAPSCORE. Site to identify the names of genes and proteins. `http://bionlp.stanford.edu/gapscore/`.
38. H.P Gardner, G.B. Wertheim, S.I. Ha, N.G. Copeland, D.J. Gilbert, and N.A. Jenkins et al. Cloning and characterization of hunk, a novel mammalian snf1-related protein kinase. *Genomics*, 63(1):46–59, 2000.
39. Genenames. Clever gene names website. `http://tinman.vetmed.helsinki.fi/eng/intro.html`.
40. GENIA. Automatic information extraction from molecular biology texts. `http://www-tsujii.is.s.u-tokyo.ac.jp/GENIA/`.
41. A.P. Gilmore and L.H. Romer.
42. D. Gusfield. *Algorithms on strings, trees, and sequences: computer science and computational biology*. Cambridge University Press, 1997.
43. S. Havre, E. Hetzler, P. Whitney, and L. Nowell. Themeriver: visualizing thematic changes in large document collections. *IEEE transactions on visualization and computer graphics*, 8(1):9–20, 2002.
44. M.A. Hearst. User interfaces and visualization. In In Baeza-Yates and Ribeiro-Neto, editors, *Modern Information Retrieval*, pages 257–324. ACM Press, 1999.
45. W. Hersh. *Information retrieval: a health and biomedical perspective*. Springer Verlag, 2nd ed edition, 2002.
46. W. Hersh, S. Price, and L. Donohoe. Assessing thesaurus-based query expansion using the umls metathesaurus. In *Proc AMIA Symp 2000*, pages 344–348, 2000.
47. W.R. Hersh and R.T. Bhupatiraju. Trec genomics track overview. In *The Twelfth Text Retrieval Conference—TREC 2003*, 2003.

48. L. Hirschman, editor. *Linking biological literature, ontologies and databases: tools for users*. Association for Computational Linguistics, 2004.
49. L. Hirschman, A.A. Morgan, and A. Yeh. Rutabaga by any other name: extracting biological names. *Journal of Biomedical Informatics*, 35:247–259, 2002.
50. P. Hoffman. *Perl for Dummies*. For Dummies, 4th ed edition, 2003.
51. HUGO. The download site of the human genome organisation. `http://www.gene.ucl.ac.uk/nomenclature/code/ftpaccess.html`.
52. Hunter. Entity identification test suite generation site. `http://compbio.uchsc.edu/Hunter_lab/testing_ei/`.
53. L. Hunter and K.B. Cohen. Using ontologies for text analysis. In *Sixth annual bio-ontologies meeting Intelligent Systems for Molecular Biology*, 2003.
54. P. Jackson and I. Moulinier.
55. T.K. Jenssen, A. Laegreid, J. Komorowski, and E. Hovig. literature network of human genes for high-throughput analysis of gene expression. *Nature Genetics*, 28(1):21–28. PMID: 11326270.
56. S. Johnson. Proc. of the workshop on natural language processing in the biomedical domain. Association for Computational Linguistics, 2002.
57. D. Jurafsky and James H. Martin. *Speech and language processing: an introduction to natural language processing, computational linguistics and speech recognition*. Prentice Hall, 2000.
58. KeX. Kex download site. `http://www.hgc.ims.u-tokyo.ac.jp/service/tooldoc/KeX/intro.html`.
59. J.D. Kim, T. Ohta, Y. Tateisi, and J. Tsujii. Genia corpus—semantically annotated corpus for bio-textmining. *Bioinformatics*, 19(Suppl. 1):180–182, 2003.
60. V. Kustanovich, J. Ishii, L. Crawford, M. Yang, J.J. McGough, and J.T. McCracken et al. Transmission disequilibrium testing of dopamine-related candidate gene polymorphisms in adhd: confirmation of association of adhd with drd4 and drd5. *Molecular Psychiatry*, Molecular Psychiatry 2003. PMID 14699430.
61. W. Lehnert. Subsymbolic sentence analysis: exploiting the best of two worlds. In Barnden and Pollack, editors, *High-level connectionist models (Advances in neural and connectionist computational theory Volume 1*, 1991.
62. G. Leroy and H. Chen. Filling preposition-based templates to capture information from medical abstracts. In *Pacific Symposium on Biocomputing 2002*, pages 350–361, 2002.
63. G. Leroy, H. Chen, and J.D. Martinez. A shallow parser based on closed-class words to capture relations in biomedical text. *Journal of Biomedical Informatics*, 36:145–158, 2003.
64. D.A. Lindberg, B.L. Humphreys, and A.T. McCray. The unified medical language system. *Methods Inf Med*, 32(4):281–291, 1993.
65. LocusLink. Locuslink download site. `ftp://ftp.ncbi.nih.gov/refseq/LocusLink/`.
66. lvg. lvg (lexical variant generation) site. `http://umlslex.nlm.nih.gov/lvg/2003/index.html`.
67. D. Maglott. Locuslink: a directory of genes. In *The NCBI Handbook*, pages 19–1 to 19–16, 2002.
68. C.D. Manning and H. Schütze. *Foundations of statistical natural language processing*. MIT Press, 1999.
69. C. Martin. *Direct Memory Access Parsing*. PhD thesis, Yale University doctoral dissertation, 1991.
70. A.T. McCray. Extending a natural language parser with umls knowledge. In *Proc Annu Symp Comput Appl Med Care 1991*, pages 194–198, 1991.

71. A.T. McCray, A.R. Aronson, A.C. Browne, T.C. Rindflesch, A. Razi, and S. Srinivasan. Umls knowledge for biomedical language processing. *Bull Med Libr Assoc*, 81(2):184–194, 1993.
72. A.T. McCray, O. Bodenreider, J.D. Malley, and A.C. Browne. Evaluating umls strings for natural language processing. In *Proc. AMIA Symp. 2001*, pages 448–452, 2001.
73. A.T. McCray, A.C. Browne, and O. Bodenreider. The lexical properties of the gene ontology. In *Proc. AMIA Symp. 2002*, pages 504–508, 2002.
74. D.J. McGoldrick and L. Hunter. Descriptron: A web service for information management in the medical and biological sciences. in preparation.
75. MedMiner. `http://discover.nci.nih.gov/textmining/main.jsp`.
76. MetaMap. `http://mmtx.nlm.nih.gov/`.
77. A. Morgan, L. Hirschman, A. Yeh, and M. Colosimo. Gene name extraction using flybase resources. In *Proceedings of the ACL 2003 workshop on natural language processing in biomedicine*. Association for Computational Linguistics, 2003.
78. MUC. Message understanding conference proceedings. `http://www.itl.nist.gov/iaui/894.02/related_projects/muc/proceedings/proceed% ings_index.html`.
79. M. Narayanaswamy, K.E. Ravikumar, and K. Vijay-Shanker. A biological named entity recognizer. In *Pacific Symposium on Biocomputing 8*, pages 427–438, 2003.
80. P.V. Ogren, K.B. Cohen, G.K. Acquaah-Mensah, J. Eberlein, and L. Hunter. The compositional structure of gene ontology terms. In *Proc. of the Pacific Symposium on Biocomputing 2004*, pages 214–225, 2004.
81. T. Ohta, Y. Tateisi, J.-D. Kim, H. Mima, and J.-I. Tsujii. The genia corpus: an annotated corpus in molecular biology. In *Proceedings of the Human Language Technology Conference*, 2002.
82. D.E. Oliver, G. Bhalotia, A.S. Schwartz, R.B. Altman, and M.A. Hearst. Tools for loading medline into a local relational database. submitted.
83. T. Ono, H. Hishigaki, A. Tanigami, and T. Takagi. Extraction of information on protein-protein interactions from the biological literature. *Bioinformatics*, 17(2):155–161, 2001.
84. Porter. Porter stemmer site. `http://www.tartarus.org/~martin/PorterStemmer/`.
85. M.F. Porter. An algorithm for suffix stripping. *Program*, 14(3):130–137, 1980.
86. S. Raychaudhuri, J.T. Chang, P.D. Sutphin, and R.B. Altman. Associating genes with gene ontology codes using a maximum entropy analysis of biomedical literature. *Genome Research*, 12:203–214, 2002.
87. C.K. Riesbeck. Conceptual analyzer to direct memory access parsing: an overview. In N.E. Sharkey, editor, *Advances in cognitive science I*. Ellis Horwood Ltd, 1986.
88. E. Riloff. Automatically constructing a dictionary for information extraction tasks. In *Proc. of the eleventh national conference on artificial intelligence (AAAI-93)*, pages 811–816. AAAI Press/MIT Press, 1993.
89. T.C. Rindflesch and A.R. Aronson. resolution while mapping free text to the umls metathesaurus. In *Proc Annu Symp Comput Appl Med Care 1994*, pages 240–244, 1994.
90. T.C. Rindflesch, L. Tanabe, J.N. Weinstein, and L. Hunter. Edgar: extraction of drugs, genes, and relations from the biomedical literature. In *Pacific Symposium on Biocomputing 5*, pages 514–525, 2000.
91. W. Ruan, H. Long, D.H. Vuong, and Y. Rao. Bifocal is a downstream target of the ste20-like serine/threonine kinase misshapen in regulating photoreceptor growth cone targeting in drosophila. *Neuron*, 36(5):831–842, 2002.
92. G. Salton. *Automatic text processing: the transformation, analysis, and retrieval of information by computer*. Addison-Wesley Publishing Company, 1989.

93. R.C. Schank and R.P. Abelson. *Scripts, plans, goals, and understanding*. Halsted Press, 1976.
94. Schwartz and Hearst. Schwartz and hearst abbreviation code site. `http://biotext.berkeley.edu/software.html`.
95. A.S. Schwartz and M.A. Hearst. A simple algorithm for identifying abbreviation definitions in biomedical text. In *Pacific Symposium on Biocomputing*, pages 8:451–462, 2003.
96. D.B. Searls. The computational linguistics of biological sequences. In L. Hunter, editor, *Artificial Intelligence and Molecular Biology*, pages 47–121. MIT Press, 1993.
97. D.B. Searls. The language of genes. *Nature*, 420:211–217, 2002.
98. H. Shatkay, S. Edwards, W.J. Wilbur, and M. Boguski. Genes, themes and microarrays: using information retrieval for large-scale gene analysis. In *Proceedings of the International Conference on Intelligent Systems for Molecular Biology 2000*, pages 317–328, 2000.
99. H. Shatkay and R. Feldman. Mining the biomedical literature in the genomic era: an overview. *Journal of Computational Biology*, 10(6):821–855, 2004.
100. M. Skounakis and M. Craven. Evidence combination in biomedical natural-language processing. In *Third workshop on data mining in bioinformatics*, 2003.
101. M. Skounakis, M. Craven, and S. Ray. Hierarchical hidden markov models for information extraction. In Morgan Kaufmann, editor, *Proceedings of the 18th international joint conference on artificial intelligence*, 2003.
102. Slackware.
103. G.W. Smith. *Computers and human language*. Oxford University Press, 1991.
104. Stanford. Stanford abbreviation server. `http://bionlp.stanford.edu/abbreviation/`.
105. P.D. Stetson, S.B. Johnson, M. Scotch, and G. Hripcsak. The sublanguage of cross-coverage. In *Proceedings of the AMIA 2002 Annual Symposium*, pages 742–746, 2002.
106. L. Tanabe. *Text mining the biomedical literature for genetic knowledge*. PhD thesis, George Mason University doctoral dissertation., 2003.
107. L. Tanabe, U. Scherf, L.H. Smith, J.K. Lee, G.S. Michaels, and L. Hunter et al.
108. L. Tanabe and W.J. Wilbur. Tagging gene and protein names in biomedical text. *Bioinformatics*, 18(8):1124–1132, 2002a.
109. L. Tanabe and W.J. Wilbur. Tagging gene and protein names in full text articles. In *Proc. of the workshop on natural language processing in the biomedical domain*. Association for Computational Linguistics, 2002b.
110. ThemeRiver. Themeriver demo. `http://www.pnl.gov/infoviz/technologies.html#themeriver`.
111. TrigramsAndTags. Trigrams'n'tags (tnt) site. `http://www.coli.uni-sb.de/~thorsten/tnt/`.
112. UMLS. `http://www.nlm.nih.gov/research/umls/`.
113. C. Verspoor, C. Joslyn, and G.J. Papcun. The gene ontology as a source of lexical semantic knowledge for a biological natural language processing application. In *Participant Notebook of the ACM SIGIR'03 Workshop on Text Analysis and Search for Bioinformatics*, pages 51–56, 2003.
114. H.M. Wain, M. Lush, F. Ducluzeau, and S. Povey. Genew: the human gene nomenclature database. *Nucleic Acids Research*, 30(1):169–171, 2002.
115. A. Yakushiji, Y. Tateisi, Y. Miyao, and J. Tsujii. Event extraction from biomedical papers using a full parser. In *Pacific Symposium on Biocomputing 2001*, pages 408–419, 2001.
116. M.D. Yandell and W.H. Majoros. Genomics and natural language processing. *Nature Reviews/Genetics*, Vol. 3:601–610, Aug. 2002.

117. Yapex. Yapex data set. http://www.sics.se/humle/projects/prothalt/#data.
118. Yapex. Yapex entity identification system site. http://www.sics.se/humle/projects/prothalt/.
119. A. Yeh, L. Hirschman, and A. Morgan. Evaluation of text data mining for database curation: lessons learned from the kdd challenge cup. *Bioinformatics*, 19(Suppl. 1):i331–i339, 2003.
120. H. Yu, C. Friedman, A. Rhzetsky, and P. Kra. Representing genomic knowledge in the umls semantic network. In *Proc AMIA Symp 1999*, pages 181–185, 1999.

Systems Level Modeling of Gene Regulatory Networks

Martin Stetter[1], Bernd Schürmann[1] and Mathäus Dejori[1,2]

[1] Siemens AG, Corporate Technology, Information and Communications, Munich, Germany.
[2] Dept of Computer Science, Technical University of Munich, Garching, Germany.
Corresponding author's e-mail: stetter@siemens.com

Summary. The perhaps most important signaling network in living cells is constituted by the interactions of proteins with the genome—the gene regulatory network of the cell. From a system level point of view, the various interactions and control loops, which form a genetic network, represent the basis upon which the vast complexity and flexibility of life processes emerges. Here we review ways by which artificial intelligence approaches can help gaining a more quantitative understanding of regulatory genetic networks at the systems level.

1 Introduction

All biochemical processes in a living cell are subject to initiation and control by the genome, encoded in the deoxyribonucleic acid (DNA) molecules. For example, cells must maintain their organization, must ensure their nutrition and must synthesize and exchange new biomolecules—which they consist of —by metabolic processes. In response to external chemical signals, they eventually grow and undergo cell division, differentiate into specialized cell types, start secretion of chemical substances or initiate their own death [1]. In addition, cells must maintain their physical functionality by a variety of repair mechanisms, such as DNA repair [2] and must produce the corresponding specialized proteins selectively when they are needed.

But there are also restrictions and malfunctions of this machinery with important and often undesirable implications. For example, many severe diseases including different kinds of cancer, Alzheimer's disease and many others show a clear relationship to genetic disorders [2]. Further, fundamental restrictions in wound healing and repair exist for higher organisms (amputated limbs and removed organs do not replace themselves in humans) and seem to have their roots in a disability of differentiated cells to re-use the pluripotent genetic machinery of stem cells. Finally, changes in the cellular machinery related to impairment in repair mechanisms also seem to be involved in aging and death [3, 4].

All these processes are directly or indirectly related to and guided by complex, recurrent and mutually interacting signaling chains. Proteins are produced from genes, interact with each other and with smaller molecules but also act back onto the DNA

W. Dubitzky and F. Azuaje (eds.), Artificial Intelligence Methods and Tools for Systems Biology, 175–195.

where they regulate the production of other proteins. Hence, understanding life processes implies understanding the concerted action of large groups of biomolecules, rather than understanding their individual actions.

One major component of this intensely and recurrently interacting system is the network of regulatory relationships between genes, the gene regulatory network. Triggered by the establishment of genome-wide expression measurements with DNA microarrays, increasing efforts are being undertaken to understand how genetic networks operate at a system-wide level. Understanding genetic networks will open the gate for new in silico technologies: Technologies to help quantifying how the global collaboration of biomolecules can emerge from underlying functional principles; how these global modes of operation are controlled in a flexible way by often locally applied biosignals; how diseases can evolve as a consequence of perturbations of the equilibrium in this network; and finally, how drugs need to act such as to restore this equilibrium. We will have methods to play what-if scenarios in silico to quantify disease mechanisms, the effects of drug treatment and to support new approaches to tissue engineering.

Methods of artificial intelligence (AI) including statistical learning theory, machine learning, artificial neural networks, neurodynamical modeling and related techniques have been proven very useful tools for better understanding gene regulatory networks. Here we provide a summary over recent AI approaches towards systems level computational modeling of genetic networks. Although many methods presented are applicable to various kinds of molecular interaction networks, we will put an emphasis on the analysis of gene regulatory networks based on DNA chip measurements. Also, we will focus on methods which aim at extracting principles of cellular function from a large scale view on genetic networks. In the next section, we briefly summarize some characteristics of genetic networks and genome wide expression measurements. The following sections will summarize two different classes of genetic network models: hypothesis-driven dynamical approaches and data-driven mining methods. The chapter will be concluded by a brief summary.

2 Genetic networks and gene expression profiling

The human genome is believed to consist of about 30000 genes, which encode the sequences of about 1 million different proteins [1, 5]. When a gene is expressed, its nucleotide sequence is first transcribed into gene-specific messenger-RNA (mRNA). The mRNA is subsequently translated by ribosomes to an amino-acid chain which folds into its functional form, the protein (Fig. 1a). There are a number of modification and processing steps during transcription and translation, including RNA splicing and post-translational modifications of proteins by enzymes. In addition many proteins form various sets of aggregates, and are only operative in these complexes. Due to splicing, post-translational modifications and aggregation, each gene may produce a whole family of operative protein structures.

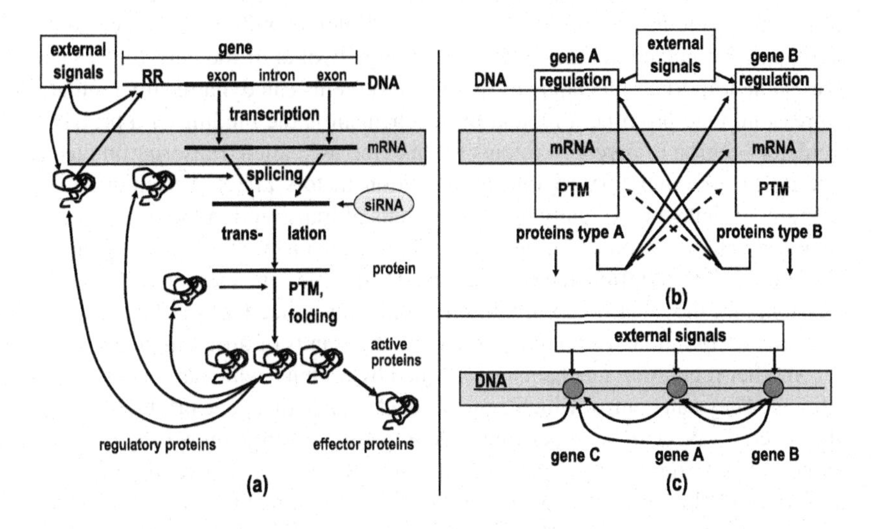

Fig. 1. From cellular regulatory mechanisms to abstract genetic networks. **(a)** Gene expression and gene regulation. Transcription and translation form proteins from genes, leaving mRNA as intermediate product. Proteins and siRNA can regulate gene expression by various mechanisms. RR= regulatory region; PTM = post-translational modification (of proteins) For details see text. **(b)** More schematic view of the interaction pathways between genes. **(c)** Abstract genetic network. Shaded boxes mark the part of the network measured by DNA microarray experiments.

2.1 Gene regulation and genetic networks

Each cell of an organism contains only a subset of all possible proteins at any time. This subset forms its proteome. Whereas the genome is (almost) the same in each cell, the proteome differs drastically for different cell types. In addition, the proteome changes dynamically depending on the state of the cell (e.g., the phase of the cell-division cycle) and on the external signals imposed on it. This implies that cells have control over the use of their genome. In fact, they have the possibility to change the proteome in a very flexible way. One major mechanism for this control is the regulation of gene expression.

Gene expression is regulated in a number of different ways. Each gene is provided with a regulatory region upstream its expressed part (Fig. 1a) During cell differentiation, specific sets of genes are permanently in-activated by cytosine methylation within the regulatory region [6]. Any modification in the DNA methylation pattern causes a changed gene expression pattern and a change in the proteome. Actually, disrupted DNA methylation patterns have been frequently observed in tumor cells [7] and their analysis has proven useful for cancer classification [8].

But also after differentiation, gene expression is dynamically regulated. In fact, most of the proteins interact with each other or interact with the genome and thereby participate in molecular signaling or reaction networks of the cell. Fig. 1a summarizes some important mechanisms, by which proteins can dynamically regulate gene expression. One prominent regulatory mechanism is the binding of a protein to the regulatory region of a gene. Proteins of this kind are called transcription factors. By binding to the regulatory region, transcription factors affect the initiation of transcription of that gene. Usually up to a few tens of transcription factors can act on the same regulatory regions of a gene [9]. This mechanism constitutes a gene-protein interaction. Transcription factors can enable, disable, enhance or repress gene expression, and they can do so in a highly nonlinear collective way [10]. In turn, any given transcription factor could act on a few thousands of different genes [5].

Another regulation mechanism is formed by proteins which are involved in RNA splicing. They can control which gene product or if at all a gene product is translated. Moreover, mRNA can be in-activated in a selective and efficient way by short double-stranded RNA fragments, a mechanism called RNA interference. Small interfering RNA (siRNA) has proven a powerful tool for investigating the effect of selective gene silencing on the organism [11]. Finally, proteins modify other proteins or attach to them. These latter mechanisms represent protein-protein interactions.

In summary, dynamic gene regulation adds a feedback step to the feed-forward process of gene expression. Gene expression controls protein concentrations, and proteins in turn—either directly or indirectly—regulate gene expression levels. Hence, the genome and proteome form a recurrent (and nonlinear) regulatory network, the gene regulatory network of the cell. The operation of the genetic network is controlled and guided by extracellular signals, which form the interface between a cell and its environment. External signals can regulate gene expression either by directly acting as transcription factors, or by modifying transcription factors [5].

Fig. 1b illustrates at a more abstract level the pathways, by which a gene A can regulate the expression level of another gene B at three different levels. Some mechanisms of regulation, including DNA methylation, transcriptional regulation, modification of transcription factors by protein-protein interactions and RNA splicing, are reflected in the cellular concentrations of mRNA (solid arrows). They can be detected in principle by gene expression measurements. Other regulatory mechanisms including RNA interference and certain types of protein-protein interaction, cannot be detected from the mRNA concentration (dashed arrows). Fig. 1c, finally sketches a fully abstracted graphical representation of a genetic regulatory network.

2.2 Gene expression measurements by DNA microarrays

Computational models of genetic networks depend on the availability of data that reflect the state of the system. These data would ideally involve the expression rates of all genes plus the state of the proteome, i.e., the types, concentrations and states of all proteins in the cell. To date, delineating the proteome as a whole is still difficult, because we lack massively parallel techniques for protein characterization. Present techniques are centered around two-dimensional gel electrophoresis followed by mass spectrometry to determine the protein sequences [5, 12], although protein chip technologies for large-scale proteome measurements are being put forward [13, 14].

During the last decade, techniques for the large scale measurement of gene expression levels, which are based on DNA microarrays and related techniques (generally referred to as DNA chips), have been developed [15, 16], for reviews see [17, 18, 19, 20]. Microarray measurements make use of the known gene sequences known from the Human Genome Project [21, 22, 23]. They are based on selective hybridization of cellular mRNA with complementary nucleotide sequences. A spot on a carrier on which such molecules with a given sequence are fixed, can therefore act as a selective probe for one type of mRNA. A DNA microarray consists of many thousands of different mRNA probes, and can therefore measure a snapshot of thousands of cellular mRNA concentrations at a time. The shaded boxes in Fig. 1 mark the level at which DNA microarray measurements operate. They make use of the fact that the presence of mRNA for a certain gene reflects the level of transcription of this gene. Therefore, DNA microarray measurements are also called gene expression profiles. It needs to be kept in mind, however, that not all regulatory processes are reflected in the mRNA concentration (counter examples include siRNA silencing and many protein-protein interactions, dashed arrows in Fig. 1b)

There are many sources of noise in microarray experiments, which include biological noise (variation of cellular states in a homogeneous strain, variations caused by RNA extraction procedures), finite sample effects such as fluctuations in the number of hybridizing molecules, optical readout noise, and others. Consequently, gene expression profiles are adequately described by probabilistic methods.

3 Dynamical models of genetic networks

Genetic regulatory networks form complex, recurrent and nonlinear systems, which evolve dynamically according to their mutual interactions and in response to external signals. Because also the brains of higher organisms represent large-scale nonlinear recurrent signal processing systems, it turns out that a large body of theoretical investigation of such systems has been accumulated by theoretical brain research. In this related field, biological networks of nerve cells have been vastly abstracted in the mid eighties to various kinds of artificial neural networks [24, 25]. During the following decade, many analogies could be found between the biologically inspired theory of neural networks and statistical learning theory [26]. This link opened the gate for the unification of machine learning techniques and neural networks and at the same time drew the attention of theoreticians towards the applicability of machine learning methods for the large-scale analysis of biological systems. Hence in brain research, an apparent step backwards, namely to more abstract artificial neural networks, has brought qualitatively new insights into the systems biology of the brain, based on which —as a second step—more biological knowledge and diversity is nowadays taken into account in the younger discipline of computational neuroscience [27, 28] to provide more detailed brain models.

At present, many approaches towards genome research seem to be at the edge of the first abstraction step. Although first attempts to describe genetic networks at an abstract level date back more than thirty years [29, 30], the data basis to constrain the models has been too sparse at that time. However, as described in the previous section, increasing knowledge about gene sequences and a growing body of DNA microarray technologies and data sets now open the gate for a new generation of computational large-scale analysis [31] and modeling [32] methods.

3.1 Biochemically inspired models

Biochemically inspired models of genetic networks are based on the reaction kinetics between the different components of the genetic regulatory network within the different compartments of a cell. These models can be associated with the level of detail of Fig. 1a. Reaction kinetics provides a framework, by which the chemical reactions between molecular compounds of the network can be described [33, 34]. For example, if a transcription factor T, expressed from a gene j, is brought together with the DNA sequence it selectively binds to, say in the regulatory region of of gene i, it might react with a rate k_1 to form a compound TD with the DNA, but might also dissociate with a rate k_2 from it. If this process follows a first order reaction kinetics, its time evolution is given by

$$\frac{d}{dt}[TD]_i(t) = -k_2[TD]_i(t) + k_1[T]_i(t); \tag{1}$$

$$\frac{d}{dt}[T]_i(t) = -k_3[T]_i(t) + k_2[TD]_i(t) + d[T_{ext}]_i(t). \tag{2}$$

In these equations, $[A]_i$ denotes the probability of finding molecule A at the site of the considered gene i. The term $d[T_{ext}]_i$ denotes the net extra supply of transcription factor molecules T at gene i. It can be thought to originate from the expression of T at gene j, followed by its diffusion towards the site of gene i. This implements a causal relationship from gene j to gene i. Molecule T can also diffuse away from i or can be degraded by chemical processes, with a rate k_3. Many gene-gene-interactions and causal relationships form the gene regulatory network described by the model. If diffusion processes across the intracellular space are explicitly taken into account, the resulting models are referred to as reaction-diffusion models. Collective phenomena in reaction-diffusion systems have been first described more than 50 years ago [35].

Biochemically inspired models have the advantage that they can be most directly related to biological processes, but they also suffer from a number of difficulties. For example, most of the biochemically relevant reactions under participation of proteins do not follow linear reaction kinetics. Many proteins undergo conformational changes after reactions, which change their chemical behavior. In particular, in many regulatory DNA regions transcription factor binding can show cooperative or competitive effects, which are nonlinear and mostly unknown. Second, the full network of metabolic, enzymatic and regulatory reactions is very complex and hard to disentangle in a single step. To do so, the kinetic equations of all the different interactions (e.g., those in Fig. 1a) would have to be written down, but the type of reactions and their parameters are often unknown. At present, the data basis seems therefore not sufficient to globally understand regulatory networks at this level of detail. However, there exist very well-examined regulatory sub-networks [36, 10, 37], which are sufficiently well-characterized to be modeled at the reaction kinetics level. Another promising line could be to infer the topology and reaction parameters of large-scale molecular networks by genetic programming [38].

Other approaches use approximations to reaction-kinetic formulations to arrive at systems of coupled differential equations for describing the time course of gene-expression levels [39, 34, 40]. In a differential equation approach, the temporal evolution of the expression level x_i of gene i is guided by the concerted influence of other gene products. The latter is described by a function $F(\mathbf{x})$, where $\mathbf{x} = (x_1, ..., x_N)$ is the vector of gene expression levels:

$$\tau \frac{dx_i}{dt} = -x_i + F_i(\mathbf{x}) \tag{3}$$

Differential equation models represent approaches, which adopt a more abstract view on genetic regulatory networks. For example, in eq. (3) no clear specification is provided (nor needed) anymore about whether the quantities x_i denote mRNA concentrations or protein concentrations. These models act at the level of description shown in Fig. 1c rather than Fig. 1a. The next paragraph provides a brief overview over different genetic network models of the type eq. (3).

3.2 Neural network models

The general form of eq. (3) represents a very rich family of networks. When modeling genetic networks, one major task is to specify the interaction functions F_i such as to be simple enough to be constrained by the existing data and at the same time accounting for biological evidence. One way to specify the interaction function is to assume that the effects of transcription factors j superimpose linearly when they bind to the regulatory region of a gene i, and that the expression of i is a nonlinear function f_i of the total regulatory input:

$$\tau \frac{dx_i}{dt} = -x_i + f_i \left(I_{ext,i} + \sum_j w_{ij} x_j \right) . \tag{4}$$

The weights w_{ij} describe the relative impact of each transcription factor and $I_{ext,i}$ represents external cellular signals. Equation (4) is identical to the formulation of a continuous artificial neural network [24, 25].

As mentioned previously, many biochemical reactions do not act linearly. This observation can be accounted for by including higher-order terms to the formulation of the total input [26]:

$$\tau \frac{dx_i}{dt} = -x_i + f_i \left(I_{ext,i} + \sum_j w_{ij} x_j + \sum_{jk} w_{ijk} x_j x_k + ... \right). \quad (5)$$

In this formulation, the functions F_i are specified as univariate nonlinear transforms of the Taylor-series expansion of their input arguments. This model accounts for multiplicative nonlinearities in the molecular interactions, which are quite common in reaction kinetics. A special case of the multiplicative network in eq. (5) has been shown to arise as an approximation from a reaction-kinetic formulation of transcriptional regulation [34]: When products of the activator genes j and inhibitor genes k bind to the regulatory region of gene i, the expression of gene i follows approximately

$$\frac{dx_i}{dt} = -ax_i + b \prod_j x_j^{h_j} - c \prod_{jk} x_j^{h_j} x_k^{h_k}, \quad (6)$$

where h^j and h^k are integer stoichiometric coefficients. They enumerate, how many protein molecules of type j or k (i.e., expressed from gene j or k, respectively) are required to complete a chemical binding reaction.

The complexity of the class of interaction functions can be also reduced by giving up their continuous nature. Gene expression states are then characterized by binary variables with the states 'expressed' or 'not expressed'. In this case, the functions F_i are binary-valued, with binary valued arguments, they are boolean functions. Boolean networks have been among the earliest approaches towards genetic network modeling [29].

The model formulations above have many free parameters. In fact, even the simplest neural network formulation eq. (4) is controlled by all the weights w_{ij}, which are of the order of the square of the number of genes, plus further free parameters describing the nonlinear functions. Hence the number of weights can well exceed one million, whereas the number of microarray measurements usually ranges between tens up to a few hundreds of measurements. Hence, at present these parameters cannot be estimated from the data for large network studies. In addition, all these model assume the structure of the genetic network as known. However, it is one of the major challenges of the post-genomic era to infer, which genes might act in a regulatory way on which other genes, in other words to infer the structure of the network from the data.

4 Data driven modeling of genetic networks

Therefore, another line of genetic network modeling is characterized by a yet higher level of abstraction, and treats the task of modeling microarray data as a data mining problem [41, 42, 43, 44, 45, 46, 47]. The goal of data mining is to explore a data set and to discover regularities and structures from it. As opposed to hypothesis-driven approaches, which search for a particular and pre-defined pattern in the data, data mining approaches specify autonomously, which patterns are present and important in the data —they are exploratory and data driven. As gene expression data sets are noisy by their nature, statistical methods play an important role for finding trends and patterns in the experimental results. In the case of genetic networks, one kind of patterns to be inferred can be for example clusters of genes, which are co-expressed when the cell is in a given state. Another type of pattern might be the structure of regulatory relationships between genes.

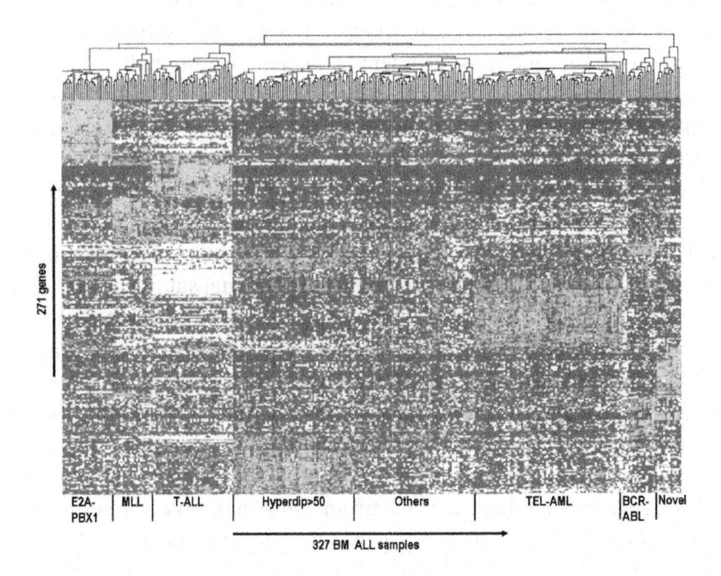

Fig. 2. Matrix of 327 gene expression profiles from ALL patients (columns) with 271 gene probes each (rows), after 2D-hierarchical clustering. White: Over-expressed state, gray: under-expressed state; black: normal expressed state with respect to a reference tissue. The genome wide expression patterns is a reliable disease marker for different ALL subtypes (names indicated below the gene expression matrix).

4.1 Clustering gene expression patterns

Clustering algorithms aim at discovering sets of genes or gene expression patterns which are more similar to each other than to others [48, 49, 50]. The set of gene expression measurements is considered as a data matrix $\mathbf{X}$, where the element x_{ij} represents the expression level of gene i in the j-th experiment [48]. Hence, each column of $\mathbf{X}$ contains the expression levels for all genes of one experiment and each row contains the expression levels of one gene across experiments. Hierarchical clustering of the rows of the matrix has been suggested in order to find clusters of genes that are co-expressed over the different experiments. One kind of hierarchical clustering groups together pairs of clusters, starting out from the individual data vectors, and proceeding along decreasing similarity [51, 50, 52].

Here we exemplify a typical application of clustering: the prediction of different subtypes of childhood acute lymphoblastic leukemia (ALL) from DNA microarray profiles measured and published by Yeoh and coworkers [52, 53]. ALL is a heterogeneous disease. Leukemic ALL-cells are related to bone marrow cells, which are destined to either become T-lymphocytes (T-lineage) or B-lymphocytes (B-lineage). The disease appears in various subtypes, which differ markedly in their response to chemotherapy. Therefore it is important to know which subtype a patient suffers from in order to design the most efficient treatment. Traditionally, the identification of the ALL subtype was a difficult and expensive process which required a combination of laboratory studies including immunophenotyping, cytogenesis and molecular diagnostics.

Recently, it has been found that the genome-wide expression pattern is a very efficient disease marker for the different ALL subtypes [52]. Fig. 2 shows a large data set of 327 microarray measurements taken from different ALL patients. 271 genes were selected as the ones which vary most with individual subtypes. The figure shows the order of the patients after hierarchical clustering of the column vectors. It can be seen that there are sharp transitions between markedly different global expression patterns. Each of these global patterns can be assigned to one disease subtype (indicated below the data set) with very high reliability. Hence, genome wide expression profiles are very efficient markers for individual disease subtypes.

Clustering can also be carried out along all two dimensions. Comparing different rows in Fig. 2 shows, that there are also groups of genes which are co-expressed, so-called gene clusters. These clusters of genes might carry out some concerted global action, which contributes crucially to the pathogenesis of an ALL subtype. In each subtype another gene cluster or set of clusters is recruited, which means that the disease mechanisms for the various ALL subtypes are probably radically different from each other.

4.2 Graphical modeling of genetic network structures

Clustering studies have revealed several large groups of genes, which collectively change their expression levels when a cell or tissue changes from one mode of life to another. These global patterns are thought to reflect the execution of specific genetic programs. One might ask the question which are the mechanisms that either stabilize a genetic program or or evoke a change to a new program. In other words, are there dominant genes or small groups of genes which are the underlying cause of a specific global gene-expression pattern? The ability to form the link between *local* causes and *global* collective behavior might provide key technologies for playing what-if scenarios in silico. These technologies will help to better understand principles of cellular function, disease mechanisms, and drug treatment on a more quantitative basis.

Due to these considerations, recent data-driven approaches have increasingly concentrated on inferring the structure of the underlying genetic regulatory networks from the statistics of microarray data, by use of graphical models [54, 55, 56, 57, 58, 59]. These approaches assume, that underlying biological interactions, such as for example transcription factor binding, will cause characteristic co-expression patterns for different genes. More generally speaking, existing biological relationships between genes or their mRNA will be more or less directly reflected in statistical relationships in the gene expression profiles. Hence, modeling the multivariate statistics of the gene expression profiles will help characterizing certain features of the underlying biological relationships causing them.

One frequently used type of graphical model are *Bayesian networks*. A Bayesian network B is a probabilistic model which describes the multivariate probability distribution for a set of variables $\mathbf{x} = \{x_1, ..., x_N\}$, where each variable x_i only depends on its parents Pa_i:

$$P(x_1, ..., x_N) = \prod_{i=1}^{N} P(x_i | Pa_i) \tag{7}$$

The associations among the variables, namely the conditional dependencies and independencies, are described by means of a directed acyclic graph (DAG) G. In the context of genetic pathway inference, each node of a Bayesian network is assigned to a gene, and can assume the different expression levels of this gene throughout the set of measurements. Each edge between genes describes a statistical dependency between them. In general, there is no single and unique graph structure for a given probability distribution. Instead, a whole set of graph structures, called an *equivalence class*, is uniquely defined by the probability distribution. If the direction of an edge between two genes is uniquely defined throughout the equivalence class, it can be interpreted as a causal relationship: It provides an estimate which gene controls another gene.

Learning Bayesian networks [60, 61, 62] use Bayesian statistics to find the network structure and the corresponding model parameters which describe best the probability distribution from which the data set $\mathbf{X}$ is drawn. In the class of score-based learning algorithms, the goodness of fit of a network G with respect to the data set $\mathbf{X}$ is assessed by assigning a score $S(G)$ by use of a statistically motivated scoring function S, as for example the *Bayesian score* [63]. It is proportional to the posterior probability $P(G|\mathbf{X})$ of the graph structure given the gene expression data matrix $\mathbf{X}$:

$$S(G|\mathbf{X}) = \frac{P(\mathbf{X}|G)P(G)}{P(\mathbf{X})}. \tag{8}$$

Unfortunately the task of structure learning is NP-hard. Moreover, usual data sets are very high-dimensional (hundreds to thousands of dimensions) with only few data points (tens to hundreds of data points). Hence, at first sight any estimate of graph structures from such data seems to suffer from an extreme lack of robustness.

A recent study has therefore focused on the problem to specify, which properties of a learned graph structure are robustly detected, when a structure with many nodes and edges must be learned with less and less data points [57]. For this, data sets with different sample sizes were drawn from the ALARM network [64]. The ALARM network is a medical diagnostic system for patient monitoring. It has 37 nodes which

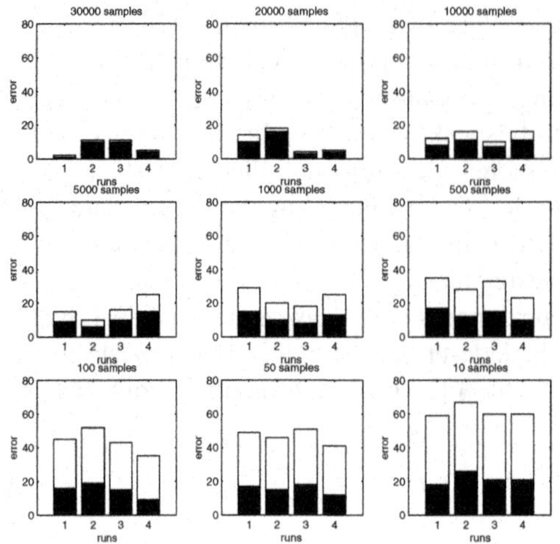

Fig. 3. Number of false positive (black bar) and false negative (white bars) edges of learned structures with respect to the true structure of the Alarm network, as a function of the size of the data sample. Each plot shows the result of ten simulation runs. From top left to bottom right, the number of data points was 30 000, 20 000, 10 000, 5 000, 1 000, 500, 100, 50, 10.

stand for medical state variables (8 diagnoses, 16 findings and 13 intermediate variables). Variables assume between 2 and 4 values. The nodes are connected by 46 directed edges, which can be interpreted in a causal manner.

Fig. 3 summarizes the number of false positive (black bars) and false negative edges (white bars) in the learned structure with respect to the true structure. Each plot summarizes four runs, and the different plots correspond to different sample sizes. The errors increase as soon as the sample size shrinks below 10000. However, it can also be observed that almost exclusively the number of false negative edges increases drastically as the sample size goes down, whereas the false positive errors rise only moderately. This behavior has an important implication for pathway inference by Bayesian networks: A Bayesian network applied to a small gene expression data set will most likely not detect all gene interactions which are actually generated by the underlying gene regulatory network. However the ones that are detected are relatively reliable even when the data set is small.

In light of these results it seems not unrealistic to assume, that the edges found by a learning Bayesian network represent real effects of the underlying genetic regulatory network rather than statistical fluctuations. Accordingly, a number of studies have suggested a structural analysis of learned graph structures [54, 57, 58]. Fig. 4a shows a part of an estimated regulatory network trained using the same ALL data set as used in Fig. 2. The network has been assembled from $Q = 20$ individual Bayesian networks using a bootstrap procedure [65, 58], in order to estimate the variability of the learning results and to keep only significant edges. For the bootstrap procedure, first a set of Q replica data sets was generated by sampling **with replacement** from the original data set. Each bootstrap replica contains as many points as the initial data set, however some points might be present in multiple copies or might be missing. Then, Q Bayesian networks are learned. After learning, the confidence in an arbitrary network feature can be assessed empirically by calculating its mean over the bootstrap replicas. For example, the confidence of an edge between two genes being

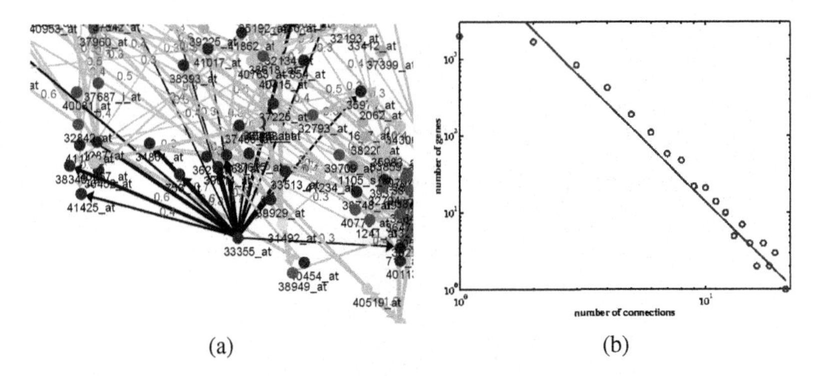

Fig. 4. **(a)** Section of the ALL network. Most genes are linked to only one or two other genes. A few genes, however, (e.g., Affymetrix-ID: 33355_at) are linked to many others. **(b)** The degree distribution of the ALL network follows a power law.

present is just the relative frequency of bootstrap networks in which this edge was present. Fig. 4a shows a part of the *feature graph* of the ALL network, in which only edges with confidence level greater than 0.5 have been included.

It can be seen that many genes maintain links to only one or two other genes, however a few of them are linked to many other genes. (e.g., gene 33355_at, Affymetrix-ID, in Fig. 4a). This observation has led to the suggestion that the densely linked genes, so-called 'dominant genes', might be of particular importance for global network functions [54]. In fact, all the dominant genes found in the ALL network are annotated either as oncogenes or as genes involved in critical cellular functions [58]. For example, the highly linked gene 33355_at mentioned above encodes the proto-oncogene *PBX1*, which is known to cause the 'E2A-PBX1' subtype of ALL after mutation.

However, there is a second observation linked with the network structure. One prominent way to quantify the link structure of a net is to calculate its degree distribution. The degree of a gene is defined as the number of links, i.e., graph edges, from and to it. Fig. 4b displays the degree distribution, i.e., the frequency of genes with link degree k. The degree distribution of the ALL network follows a power law: $P(k) = p_0 k^{-\gamma}$ with an exponent $\gamma = 3.2$. Networks with a power-law degree distribution are called *scale-free networks* [66]. Scale-free networks share certain features regarding the performance: The network operation is very robust against the damage of a random node. In contrast, a targeted attack to certain few nodes can cause a global breakdown of the whole network. In light of this observation and the scale-free property of the ALL network, it has been recently formulated a new criterion for identifying globally important genes: Genes which represent spots of high vulnerability of the network, are ranked as having high importance for the global network operation. Besides genes with high degree, also genes with a high traffic load can be identified as spots of high vulnerability. And in fact, all spots of high vulnerability of the ALL network are either known as oncogenes or are involved in

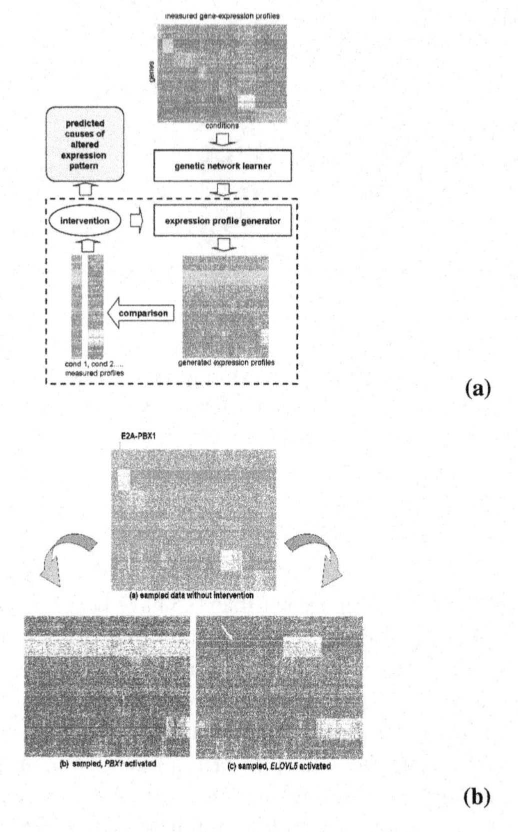

Fig. 5. **(a)** Principle of generative inverse modeling. For details see text. **(b)** Inverse modeling for ALL data after clustering. Top: Artificially generated expression profiles without intervention. Arrow marks the cluster of E2A-PBX1 characteristic profiles. These profiles are closest to the measured profiles from E2A-PBX1 patients. Bottom left: generated profiles, when gene PBX1 is kept active. Bottom right: generated profiles, when ELOV5 is activated (arrow: row of constantly over-expressed ELOV5). PBX1 stabilizes the pathological pattern more than ELOV5. Gray scale as in Fig. 2.

DNA repair, induced cell death (apoptosis), cell-cycle regulation or in other critical processes. [58]. Genes are often co-expressed and become active as a whole group, a gene cluster (Fig. 2). Further, many proteins consist of several subunits, encoded by different genes, and become only functional when the corresponding genes are all expressed at the same time. Moreover, proteins can form larger assemblies and carry out their function only in this formation. Finally, proteins can also be part of reaction cascades which subserve a common task. In summary, gene products are often naturally linked to functional modules by the one or the other of these mechanisms [67]. Hence, expressing genes often means expressing groups of genes whose products are necessary to build an operative functional module. In light of the ubiquity of modular substructures in regulatory networks, another body of work has designed machine learning methods which are tailored to describe and learn modular structures. Typical examples include module networks [68] and decomposable models [69, 59].

4.3 Generative inverse modeling of genetic networks

Structural considerations regarding graphical models of genetic regulatory networks can be complemented by the view of Bayesian networks as generative models. Instead of analyzing explicitly the graph structure, the Bayesian network is used to generate artificial gene expression patterns. If in the previous section the graph structure was related to the *structure* of the underlying genetic network, generative modeling relates the *function* of the underlying genetic network to features of the gene expression profiles.

Recently, a method called *generative inverse modeling* has been formulated, which allows to generate gene expression profiles from the Bayesian network, while certain interventions are imposed onto the model network [70]. Fig. 5a sketches the procedure of generative inverse modeling. First, a set of Bayesian networks is learned from a measured gene expression data set. After learning, the statistical relationships between gene expression levels are imprinted in the structure and parameters of the Bayesian networks. Learning several networks following a bootstrap procedure helps assessing the reliability of various network features.

After learning, different sets of artificial gene expression patterns are generated, each under the action of a different intervention E. For example, an intervention can consist of keeping fixed a gene or a set of genes at the over-expressed or under-expressed levels. Keeping fixed such individual gene expression levels could model the effect of a drug, a genetic mutation or other mechanisms related to disease generation and prevention. Fig. 6 illustrates the procedure of data sampling. By drawing gene expression profiles under interventions is then possible which effect on the *system wide* behavior of the network in terms of gene expression is evoked by this *local* interventions. Artificial gene expression patterns can then be compared to different measured ones, say, from healthy subjects and from patients suffering from different ALL subtypes (Fig. 5a , dashed box). If an artificial expression pattern under an intervention E becomes very similar to a certain pathological expression pattern, this intervention is likely to be critically involved in the pathogenesis of this disease subtype.

Fig. 5b shows an example of generative inverse modeling based on the ALL data set. On top is shown a set of 327 artificial gene expression profiles drawn from the original learned bootstrap-weighted Bayesian network. Similar to the measured data set (Fig. 2) , the profiles cluster into different patterns characteristic for ALL subtypes. When gene *PBX1* is kept over-expressed throughout data sampling, the global pattern characteristic for the E2A-PBX1 subtype of ALL is strongly stabilized. This can be seen by comparing the most frequent patterns in the bottom-left data matrix with the global shape of the E2A-PBX1-specific pattern in the measured data. In contrast, when another gene within the same gene cluster as *PBX1* is kept over-expressed (gene *ELOV5*, intervention E_2), the global pattern is only weakly stabilized. It can be

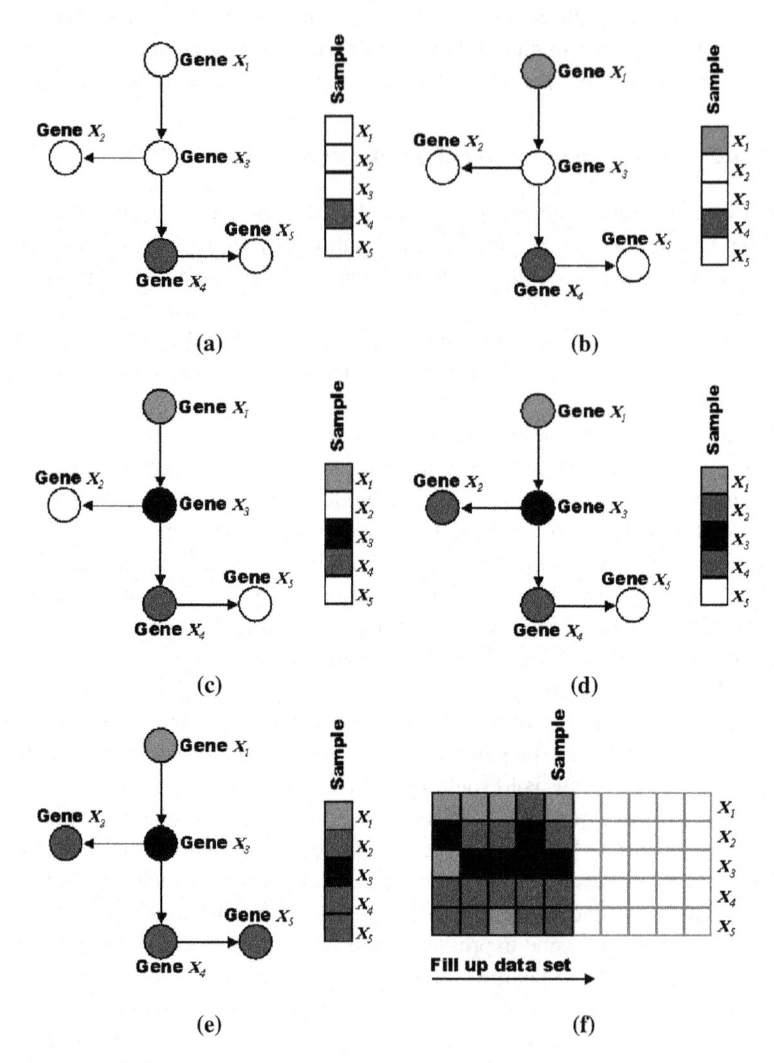

Fig. 6. Principle of the generation of artificial gene expression patterns for an example network of only five genes. **(a)** As an intervention, gene X_4 is clamped to $X_4 = +1$. **(b)-(e)** Patterns are drawn as random samples from the multivariate conditional probability distribution resulting from the intervention, by repeated Bayesian inference. For this, genes are first sorted such that parents precede children. Then, the network is filled from above: **(b)** Instantiate gene $X_1 = x_1$ according $P(X_1|X_4 = 1)$ (e.g., $X_1 = -1$). $P(X_1|X_4 = 1)$ is calculated by applying Bayes' law and subsequently summing over all other free variables. **(c)-(e)** Repeat the component-wise sampling following the node order. **(f)** Proceed with steps **(b)-(e)** until the desired data set is drawn.

concluded, that *PBX1* is more critical for the pathogenesis of this ALL subtype than *ELOV5*. In fact, *PBX1* is known as oncogene of the E2A-PBX1 disease subtype: The origin of E2A-PBX1 ALL has been correctly identified by generative inverse modeling.

In summary, one particular strength of Bayesian networks and other data driven modeling techniques is to make efficient use of the restricted data basis for obtaining information about genetic networks at the systems level. Operating at this abstract level, these approaches could well be used also to analyze protein and other molecular interaction networks, provided that there exist system-wide measurement techniques. However this generality comes at the expense of detail in the description of biochemical processes. By their philosophy, graphical modeling approaches focus only on the statistical nature of the data and ignore real biochemical processes as well as dynamical features of molecular interactions in the cell, which underly and generate gene expression patterns. We are looking forward to exciting new and probably hybrid methods to be developed, which span the whole variety of cellular phenomena to be characterized.

5 Conclusions

The genes of a genome are expressed to form the cell's proteome, which in turn acts back onto the genome in a multitude of regulatory processes. Characterizing the global principles of operation of these genetic regulatory networks represents one of the major challenges of the post-genomic era. Understanding genetic networks will help opening the gate towards a quantitative understanding of morphogenesis and pathogenesis and towards the development of new tissue engineering techniques and drug discovery methods, just to mention a few. With high-throughput gene-expression profiling techniques such as DNA microarrays, a data basis has become available for data-driven modeling of genetic networks. We have provided an overview over different approaches from artificial intelligence towards genetic network modeling. These techniques involved artificial neural networks, machine learning and data mining techniques.

To date, clustering algorithms and graphical models seem to be highly important approaches for modeling the currently available data basis of DNA microarray measurements. Learning and analyzing statistical structures of graphical models from sets of microarray data - guided and constrained by various sources of prior knowledge - can form the basis of highly useful and powerful support-systems for life sciences. These systems will allow to play what-if scenarios of various cellular interventions in silico and will put forward entirely new approaches towards understanding normal life processes and disease mechanisms.

Acknowledgments

Mathäus Dejori gratefully acknowledges support through the Ernst-von-Siemens foundation.

References

1. H. Lodish, A. Berk, S. L. Zipursky, P. Matsudaira, D. Baltimore, and J. Darnell. *Molecular Cell Biology*. W. H. Freeman and Company, 2000.
2. V. A. Bohr. DNA damage and its processing, relation to human disease. *J. Inherit. Metab. Dis*, 25(3):215–222, 2002.
3. Dana Pe er. Deletional mutations are the basic cause of ageing: historical perspectives. *Mutat. Res*, 338(1-6):3–17, 1995.
4. J. Vijg and M. E. Dolle. Large genome rearrangements as primary cause of ageing. *Mech Ageing Dev*, 123(8):907–915, 2002.
5. T. A. Brown. *Genomes*. Bios Scientific Publishers, Oxford, 1999.
6. K. D. Robertson and P. A. Jones. DNA methylation: past, present and future directions. *Carcinogenesis*, 21:461–467, 2000.
7. C. Plass. Cancer epigenomics. *Human Mol. Genomics*, 11:2479–2488, 2002.
8. F. Model, P. Adorjan, A. Olek, and C. Piepenbrock. Feature selection for DNA methylation based cancer classification. *Bioinformatics*, 17:S157–S164, 2001.
9. D. S. Latchman. *Eukaryotic transcription factors. 3rd edition.* Academic Press, San Diego, CA, 1998.
10. C. H. Yuh, H. Bolouri, and E. H. Davidson. *Cis*-regulatory logic in the *endo* 16 gene: switching from a specification to a differentiation mode of control. *Development*, 128:617–629, 2001.
11. M. Scherr, M. A. Morgan, and M. Eder. Gene silencing mediated by small interfering RNAs in mammalian cells. *Curr. Med. Chem.*, 10:245–256, 2003.
12. JR Yates. Mass spectrometry and the age of the proteome. *J. Mass Specrom.*, 33:1–19, 1998.
13. E. Phizicky, P. I. Bastiaens, H. Zhu, M. Snyder, and S. Fields. Protein analysis on a proteomic scale. *Nature*, 422(6928):208–215, 2003.
14. S. Hanash. Disease proteomics. *Nature*, 422(6928):226–232, 2003.
15. S. P. Fodor, R. P. Rava, X. C. Huang, A. C. Pease, C. P. Holmes, and C. L. Adams. Multiplexed biochemical assays with biological chips. *Nature*, 364:555–556, 1993.
16. M. Schena, D. Shalon, R. Davis, and P. O. Brown. Quantitative monitoring of gene-expression patterns with a cDNA microarray. *Science*, 270:467–470, 1995.
17. P. O. Brown and D. Botstein. Exploring the new world of the genome with DNA microarrays. *Nature Genetics*, 21:33–37, 1999.
18. M. Schena. *Microarray biochip technology*. Eaton Publishing Co., Natick, 2000.
19. P. Baldi and G. W. Hatfield. *DNA microarrays and gene expression*. Cambridge university press, Cambridge MA, 2002.
20. M. Stetter, G. Deco, and M. Dejori. Large-scale computational modeling of genetic regulatory networks. *Artificial Intelligence Review*, 20:75–93, 2003.
21. F. R. Blattner, G. Plunkett, C. A. Bloch, N. T. Perna, and V. Burland et al. The complete genome sequence of *escherichia coli* k-12. *Science*, 277(5331):1453–1474, 1997.
22. A. Goffeau, BG Barrel, R. W. Davis, B. Dujon, and H. Feldman et al. Life with 6000 genes. *Science*, 274(5287):546,563–567, 1996.
23. http://www.genome.gov, 2003.
24. J. Hertz, A. Krogh, and R. G. Palmer. *Introduction to the theory of neural computation*. Addison-Wesley, Redwood City CA, 1991.
25. S. Haykin. *Neural Networks. A comprehensive foundation*. MacMillan College Publishing Company, New York, Toronto, Oxford, 1994.
26. C. M. Bishop. *Neural Networks for pattern recognition*. Clarendon press Oxford, 1995.

27. E. T. Rolls and G. Deco. *Computational Neuroscience of Vision.* Oxford University Press, Oxford, 2002.
28. M. Stetter. *Exploration of cortical function.* Kluwer academic publishers, Dordrecht, 2002.
29. S. A. Kauffman. Homeostasis and differentiation in random genetic control networks. *Nature*, 224:177–178, 1969.
30. S. A. Kauffman. Gene regulation networks: a theory for their global structures and behaviors. In A. N. Other, editor, *Current topics in developmental biology*, volume 6, pages 145–182. Academic Press, New York, 1977.
31. D. Berrar, W. Dubitzky, and M. Granzow, editors. *A practical approach to microarray data analysis.* Kluwer Academic Publishers, Boston Dordrecht London, 2002.
32. D. Endy and R. Brent. Modelling cellular behaviour. *Nature*, 409:391–395, 2001.
33. A. Cornish-Bowden. *Fundamentals of enzyme kinetics.* Portland Press, Colchester, 1995.
34. M. Kato, T. Tsunoda, and T. Takagi. Inferring genetic networks from dna microarray data by multiple regression analysis. *Genome informatics*, 11:118–128, 2000.
35. A. M. Turing. The chemical basis of morphogenesis. *Phil. Trans. R. Soc. Lond. B*, 237:37–72, 1951.
36. C. H. Yuh, H. Bolouri, and E. H. Davidson. Genomic *cis*-regulatory logic: experimental and computational analysis of a sea urchin gene. *Science*, 279:1896–1902, 1998.
37. H. de Jong, M. Page, C. Hernandez, and J. Geiselmann. Qualitative simulation of genetic regulatory networks: method and application. In B. Nebel, editor, *Proceedings of the 17. International Joint Conference on Artificial Intelligence IJCAI-01*, pages 67–73. Morgan Kauffman, San Mateo CA, 2001.
38. J. R. Koza, W. Mydlowec, G. Lanza, J. Yu, and M. A. Keane. Reverse engineering of metabolic pathways from observed data using genetic programming. In *Proceedings of the 2001 Pacific Symposium on Biocomputing*, pages 434–445, 2001.
39. T. Chen, H. L. Le, and G. M. Church. Modeling gene expression with differential equations. *Proc. Pacific Symp. Biocomputing*, 4:29–40, 1999.
40. E. Sakamoto and H. Iba. Inferring a system of differential equations for a gene regulatory network by using genetic programming. *Congress on evolutionary computation*, pages 720–726, 2001.
41. R. Somogyi, S. Fuhrman, M. Askenazi, and A. Wuensche. The gene expression matrix: towards the extraction of genetic network architectures. In *Nonlinear Analysis, Proc. of the Second World Congress of Nonlinear Analysis (WCNA96)*, volume 30, pages 1815–1824, 1997.
42. P. D'haeseleer, X. Wen, S. Fuhrman, and R. Somogyi. Mining the gene expression matrix: inferring gene relationships from large-scale expression data. In M. Holcombe and R. Paton, editors, *Information processing in cells and tissues*, pages 203–212. Plenum Press, 1997.
43. T. J. L. Wang, B. A. Shapiro, and D. Shasha. *Pattern Discovery in Biomolecular Data: Tools, Techniques and Applications.* Oxford University Press, Oxford, 1999.
44. P. Baldi and S. Brunak. *Bioinformatics, the Machine Learning Approach.* MIT Press, Cambridge, MA, 1998.
45. P. D'haeseleer, S. Liang, and R. Somogyi. Genetic network inference: from co-expression clustering to reverse engineering. *Bioinformatics*, 16:707–726, 2000.
46. D. K. Slonim. From patterns to pathways: gene expression data analysis comes of age. *Nature Genetics*, 32 Suppl:502–508, 2002.
47. D. P. Berrar, C. S. Downes, and W. Dubitzky. Multiclass cancer classification using gene expression profiling and probabilistic neural networks. *Pac. Symp. Biocomput.*, pages 5–16, 2003.

48. M. B. Eisen, P. T. Spellman, P. O. Brown, and D. Botstein. Cluster analysis and display of genome-wide expression patterns. *Proc. Natl. Acad. Sci. USA*, 95(25):14863–14868, 1998.
49. A. Ben-Dor, R. Shamir, and Z. Yakhini. Clustering gene expression patterns. *J. Comput. Biol.*, 6(3-4):281–297, 1999.
50. T. R. Golub, D. K. Slonim, P. Tamayo, M. Gaasenbeek, J. P Mesirov, and H. Coller et al. Molecular classification of cancer: class discovery and class prediction by gene-expression monitoring. *Science*, 286(5439):531–537, 1999.
51. P. T. Spellman, G. Sherlock, M. Q. Zhang, V. R. Iyer, K. Anders, M. B. Eisen, P. O. Brown, D. Botstein, and B. Futcher. Comprehensive identification of cell cycle-regulated genes of the yeast *saccharomyces cerevisiae* by microarray hybridization. *Molecular Biology of the Cell*, 9:3273–3297, 1998.
52. E.-J. Yeoh, M. E. Ross, S. A. Shurtleff, W. K. Williams, and D. Patel et al. Classification, subtype discovery, and prediction of outcome in pediatric acute lymphoblastic leukemia by gene expression profiling. *Cancer cell*, 1:133–143, 2002.
53. http://www.stjuderesearch.org/ALL1/, 2002.
54. N. Friedman, M. Linial, I. Nachman, and D. Pe'er. Using Bayesian network analyze expression data. *J. Comput. Biology*, 7:601–620, 2000.
55. S. Imoto, T. Goto, and S. Miyano. Estimation of genetic networks and functional structures between genes by using Bayesian netork and non-parametric regression. In *Pacific Symposium on Biocomputing*, pages 175–186, 2002.
56. Seiya Imoto, T. Higuchi, T. Goto, K. Tashiro, S. Kuhara, and S. Miyano. Combining microarrays and biological knowledge for estimating gene networks via bayesian networks. In *Proceedings of the 2nd Computational Systems Bioinformatics*, pages 104–113, 2003.
57. M. Dejori and M. Stetter. Bayesian inference of genetic networks from gene-expression data: convergence and reliability. In *Proceedings of the 2003 International Conference on Artificial Intelligence (IC-AI'03)*, pages 323–327, 2003.
58. M. Dejori, B. Schürmann, and M. Stetter. Hunting drug targets by systems-level modeling of gene expression profiles. *IEEE Trans. Nano-Bioscience*, in press, 2004.
59. M. Dejori, A. Schwaighofer, V. Tresp, and M. Stetter. Mining functional modules in genetic networks by decomposable models. *OMICS*, in press, 2004.
60. R. Hofmann. *Lernen der Struktur nichtlinearer Abhängigkeiten mit graphischen Modellen.* Doctoral Thesis, 2000.
61. H. Steck. *Constraint-based structural learning in Bayesian networks using finite data sets.* PhD thesis, Doctoral thesis, Munich, Germany, 2001.
62. D. M. Chickering. *Learning Bayesian Networks from Data.* UCLA Cognitive Systems Laboratory, Technical Report (R-245), June 1996. Ph.D. Thesis., 1996.
63. D. Heckerman, D. Geiger, and D. Chickering. Learning Bayesian networks: The combination of knowledge and statistical data. *Machine Learning*, 20:197–243, 1995.
64. Norsys Software Corp. Alarm network.
http://www.norsys.com/netlib/, 2003.
65. N. Friedman, M. Goldszmidt, and A. Wyner. Data analysis with bayesian networks: a bootstrap approach. pages 196–205, 1999.
66. A. E. Motter, T. Nishikawa, and Y.-C. Lai. Range-based attacks on links in scale-free networks: are long-range links responsible for the small-world phenomenon? *Phys. Rev. E*, 66:065103, 2002.
67. L. H. Hartwell, J. J. Hopfield, S. Leibler, and A. W. Murray. From molecular to modular cell biology. *Nature*, 402:C47, 1999.

68. E. Segal, M. Shapira, A. Regev, D. Pe'er, D. Botstein, D. Koller, and N. Friedman. Module networks: identifying regulatory modules and their condition-specific regulators from gene-expression data. *Nat. Genet.*, 34(2):166–176, 2003.
69. A. Schwaighofer, V. Tresp, M. Dejori, and M. Stetter. Structure learning for nonparametric decomposable models. *J. Machine Learning Research*, submitted, 2004.
70. M. Dejori and M. Stetter. Identifying interventional and pathogenic mechanisms by generative inverse modeling of gene expression profiles. *J. Comput. Biology*, in press, 2004.

Computational Neuroscience for Cognitive Brain Functions

Marco Loh[1], Miruna Szabo[2], Rita Almeida[1], Martin Stetter[2] and Gustavo Deco[13]

[1] Dept. of Technology, Universitat Pompeu Fabra, Barcelona, Spain.
[2] Siemens AG, Corporate Technology, Information and Communications, Munich, Germany.
[3] Institució Catalana de Recerca i Estudis Avançats (ICREA), Spain.
Corresponding author's e-mail: `loh@in.tum.de`

Summary. Cognitive behavior requires complex context-dependent processing of information that partially emerges from the links between attentional perceptual processes and working memory. We describe a computational neuroscience theoretical framework which shows how an attentional bias can influence perceptual processing, the mapping of sensory inputs to motor output and formation of selective working memory. This theoretical framework incorporates spiking and synaptic dynamics which enable single neuron responses, *functional magnetic resonance imaging* (fMRI) activations, psychophysical results, the effects of pharmacological agents, and the effects of damage to parts of the system, to be explicitly simulated and predicted. This computational neuroscience framework provides an approach for integrating different levels of investigation of brain function, and for understanding the relations between them.

1 Introduction

Nowadays, the scientific community agrees that the processing of information by the brain underlies sensory, motor and cognitive functions. Neurons are the cells responsible for this processing of information, that is for the coding, transmission, and integration of signals originating inside or outside the nervous system. At least in the mammalian brain, it is believed that brain functions are achieved by the conjoint information processing of large groups of neurons. The transmission of information within and between neurons involves changes in the so-called resting membrane potential, the electrical potential of the neurons at rest, when compared to the extracellular space. The inputs one neuron receives at the synapses from other neurons cause transient local changes in its resting membrane potential, called postsynaptic potentials. These changes in potential are carried out by the flux of ions between the intra- and extra-cellular space. The flux of ions is made possible through ion channels present in the membrane. The ion channels open or close depending on the membrane potential and on substances released by the neurons, named neurotransmitters, which bind to receptors on the cell's membrane. The postsynaptic potentials

W. Dubitzky and F. Azuaje (eds.), Artificial Intelligence Methods and Tools for Systems Biology, 197–215.

can either hyperpolarize or depolarize the cell. When the conjoint effect of the postsynaptic potentials at a given time reaches a certain threshold value, above the resting membrane potential, the neuron will produce an impulse of signal. The impulses of signal, called action potentials, are characterized by a certain amplitude and duration and are the units of information transmission at the inter-neuronal level. The information is thought to be coded in terms of the frequency of the action potentials, called spiking or firing rate, as well as possibly in the timing of action potentials.

One possibility to investigate the biological basis of the processing of information in the brain is to study the response of neurons to stimulation. This can be done in experimental animals using implanted electrodes to record the rates and timing of action potentials. However, this approach is generally not possible to apply in humans. To study brain function in humans, techniques allowing the indirect study of the activity of neurons have been developed. An example of one such technique is fMRI, measuring regional changes in metabolism and blood flow, indirectly associated with regional changes in brain activity. This approach of measuring *regional* differences in brain activity is possible, because at a macroscopic level the brain, and the cortex in particular, is organized into spatially segregated regions known to have functionally specialized roles. For example, the occipital cortex is specialized for vision, with different macroscopic subregions specialized in particular types of visual information processing. Further within some of these regions there is a topographic organization, that is adjacent neurons are responsive to adjacent portions of the image seen. Topographic organizations are found in other sensory and motor systems. A technique such as fMRI allows the mapping of brain regions associated with a particular task or task component. For a detailed introductory description of neuronal physiology, cortical organization and measuring techniques see any neuroscience textbook, for example [20].

Understanding the fundamental principles underlying higher brain functions requires the integration of different levels of experimental investigation in cognitive neuroscience (from the operation of single neurons and neuroanatomy, neurophysiology, neuroimaging and neuropsychology to behavior) via a unifying theoretical framework that captures the neural dynamics inherent in the computation of cognitive processes. A theoretical framework that fulfills these requirements can be obtained by developing explicit mathematical neurodynamical models of brain function based at the level of neuronal spiking and synaptic activity [29, 8].

In this chapter, we review, using a computational neuroscience perspective, neuronal and cortical mechanisms for the interplay between visual attention and working memory, as fundamental basic processes involved in cognition. In highly developed animals, like primates, expansion of the sensory and motor capabilities increases the problem of stimulus selection, i.e., the question as to which information is relevant to react to. Paradoxically, in spite of the massively parallel character of computations performed by the brain, it seems that biological systems employ a selection processing strategy for managing the enormous amount of information resulting from their interaction with the environment. This selection of relevant information is referred to as visual attention. The concept of attention implies that we can concentrate on certain portions of the sensory input, motor programs, memory contents, or internal representations to be processed preferentially, shifting the processing focus from one location to another or from one object to another in a serial fashion. Focusing attention is dependent on the context of the task one aims at solving at present. At any given moment the context information can change and must then be updated and maintained in the mind. This process of actively storing and manipulating information in mind for a short period of time has been defined as working memory [2].

In this chapter, neurodynamics, described at the level of spiking and synaptic activity, are used to provide a quantitative formulation for the dynamical evolution of single neurons, neural networks, and coupled hierarchical modules of networks. The structure we use is organized within the general framework of the biased competition hypothesis for selective attention. [24, 32, 25, 23, 7, 27, 6], which assumes that populations or pools of activated neurons engage in competitive interactions mediated by global inhibition and that this competition can be biased toward some given neuronal populations by an external signal representing attention or context. In a generalization of this hypothesis, some populations are thought not to compete but to cooperate, that is they are thought to mutually reinforce their repective activities. Rolls and Deco [29] introduced a theoretical framework for the neurodynamics of biased competition and cooperation, in which multiple activated populations of neurons interact with each other in a hierarchical way. Neuron populations which are combined in such a way as to model an individual brain structure (e.g., a cortical area) engage in competitive and cooperative interactions with each other, trying to represent their input in a context-dependent way. But biased competition and cooperation networks often consist of several model areas. External inputs from one model area bias the internal competition and cooperation in favor of specific neurons. By this bias, each model area forms a context or *top-down hypothesis*, in the framework of which the dynamics of the other areas are guided. In the context of visual attention, for example, the external bias can be interpreted as an attentional effect which is produced by generating signals in areas outside the visual cortical areas. These signals are then fed back to extrastriate, higher visual areas where they bias the competition in such a way that, when multiple stimuli appear in the visual field, the cells representing the attended stimulus 'win', thereby suppressing cells representing distracting stimuli [29, 13, 11, 12].

The operation of biased competition and cooperation networks in general can be imagined as a negotiation process between different specialized experts: Each model area tries to represent one aspect of the environment. These aspects can be spatial relationships, object identities, history, currently valid rules, rewarding tasks and many others. Each representation alone is insufficient to deal with the complexity of the environment. However the areas bias each other and mutually guide each other's internal dynamics until a maximally coherent state is reached, which then forms a good global representation of the environment.

Multi-areal modeling of large scale brain systems strikingly exemplifies that an understanding of higher brain functions cannot be achieved by adopting a purely reductionist view. Instead, most key features of brain operation seem to emerge from the mutual interplay of the components rather than being generated by each of the components individually. Principles of brain operation can therefore only be fully understood by a holistic, systems level quantitative treatment. It turns out that many complex phenomena in living systems, even the phenomenon of life itself, appear to draw their complexity from interactions of structured networks such as gene regulatory networks, metabolic networks, protein networks, intercellular signaling networks and immunoresponse networks, just to mention a few.

In light of this, we urgently need powerful methods to understand, model and analyze densely interacting biological networks at a systems level. Four decades of AI research have provided a large pool of methods, techniques and background knowledge required to form quantitative models of the brain and other systems. In this chapter we exemplify how AI methods from the fields of artificial neural networks, spiking neuron networks and nonlinear dynamics can be used to model and understand visual attentional and cognitive phenomena as effects of multi-areal recurrent processing. By doing so, we are on our way not only toward understanding operational principles of the brain but can also hope to learn —although at a high level— about other complex networks which altogether form living biological organisms.

However, systems biology also can provide input to AI research. Living systems have developed a highly desirable property: They can successfully navigate in and manipulate a complex, changing, potentially unpredictable and open environment. For example, bacteria can robustly deal with their chemical environment, cells of a higher organism robustly act within their molecular signaling network, and finally our brain enables us to robustly navigate in our complex natural environment. To be able to do so, living systems adopt a complex linked internal structure which largely remains to be understood, they learn from the environment, adapt to changes and contain error tolerance strategies. By modeling living systems with AI methods, we hope to be able to transfer many of these desirable features to artificial systems better than achieved so far. For example, multi-areal neurodynamical models as presented here could trigger new developments for artificial neural networks, brain-like expert and decision support systems, prognosis systems and other distributed AI approaches like mixture of experts, intelligent agents in open multi-agent systems and the suitable design of these environments.

In the next section, we will briefly summarize the procedure to set up biased competition and cooperation neurodynamical models. The following sections will summarize three examples of how neuronal responses recorded while animals underwent cognitive tasks, pop out in a natural way when modeled in the context of biased competition and cooperation. The chapter will be concluded by a brief outlook to future research directions.

2 Computational neuroscience tools

What are the theoretical tools for achieving the proper level of description of the neurodynamical mechanisms that underlie brain functions? On the one hand, the level of description should be accurate enough to allow the relevant mechanisms at the level of neurons and synapses to be properly taken into account. On the other hand, the description should be simple enough, so that we can really infer by abstraction the *relevant principles* substantiating perception and cognition. A mere reproduction of phenomena of a complex system, like the brain, just by simulating the same kind of artificial complex system, is most of the time not useful, because there are usually no explicit underlying *first principles*, and it is also unrealistic.

We assume that a proper level of description at the microscopic level is captured by the spiking and synaptic dynamics of one-compartment, point-like models of neurons, such as *integrate-and-fire* (IF) models (in particular we use the model described in [5]). An IF neuron can be described by a circuit consisting of a capacitance (the cell membrane capacitance C_m) in parallel with a resistance (the cell membrane resistance R_m) driven by input currents coming from connected neurons. When the voltage across the membrane capacitance reaches a given threshold the circuit is shunted and the neuron generates a spike which is then transmitted to other neurons. The spikes arriving to a given neuron produce post-synaptic excitatory or inhibitory potentials (through a low pass filter by the synaptic and membrane time constants) and constitute the incoming input to the neuron. The dynamics of the IF model allow the use of realistic biophysical constants (like conductances, delays, etc.) in a thorough study of the realistic time scales and firing rates involved in the evolution of the neural activity underlying cognitive processes, for comparison with experimental data. We believe that it is essential in a *biologically plausible* model that the different time scales involved are properly described, because the system that we are describing is a dynamical system that is sensitive to the underlying different spiking and synaptic time courses, and the non-linearities involved in these processes. For this reason, it is convenient to include a thorough description of the different time courses of the synaptic activity for both excitatory and inhibitory postsynaptic potentials. In the neuronal model considered here the excitatory postsynaptic currents have two components, a fast one mediated by α-amino-3-hydroxy-5-methyl-4-isoxazolepropionic acid (AMPA) receptors and a slow one mediated by N-methyl-D-aspartate (NMDA) receptors. The inhibitory currents are mediated by the neurotransmitter gamma-aminobutyric acid (GABA) (see for example [30]). A second reason why this temporally realistic and detailed level of description of synaptic activity is required, is the goal to perform simulations which can be compared with fMRI data. These involve the realistic calculation of BOLD-signals that are intrinsically linked with the synaptic dynamics, as recently found by [22]. A third reason is that one can consider the influence of neurotransmitters and pharmacological manipulations, e.g., the influence of dopamine on the NMDA and GABA receptor dynamics [38, 21], to study the effect on the global dynamics and on the related cortical functions (e.g., working memory [9], and [10]). A fourth reason for analysis at the level of spiking neurons is that the computational units of the brain are the neurons, in the sense that they transform a large set of inputs received from different neurons into an output spike train, that this is the single output signal of the neuron which is connected to other neurons, and that this is therefore the level at which the information is being transferred between the neurons, and thus at which the brain's representations and computations can be understood [29, 28].

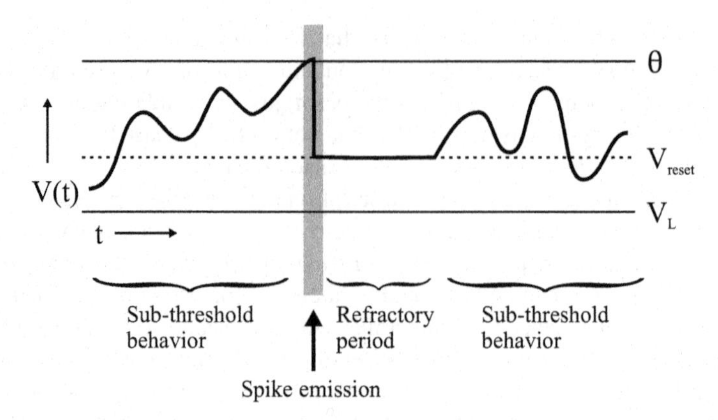

Fig. 1. Course of the membrane potential of a neuron along the time-axis. During the sub-threshold behavior the membrane potential is influence by the incoming spikes mediated by the synaptic currents $I_{\mathrm{syn}}(t)$. Once the threshold θ is reached, the neurons emits a spike and stays for a refractory period τ_{ref} at the reset potential V_{reset}

For all these reasons, the non-stationary temporal evolution of the spiking dynamics are addressed by describing each neuron by an integrate-and-fire model. The subthreshold membrane potential $V(t)$ of each neuron evolves according to the following equation:

$$C_{\mathrm{m}} \frac{dV(t)}{dt} = -g_{\mathrm{m}}(V(t) - V_L) - I_{\mathrm{syn}}(t) \tag{1}$$

where $I_{\mathrm{syn}}(t)$ is the total synaptic current flow into the cell, V_L is the resting potential, C_{m} is the membrane capacitance, and g_{m} is the membrane conductance. When the membrane potential $V(t)$ reaches the threshold θ a spike is generated, and the membrane potential is reset to V_{reset}. The neuron is unable to spike during the first τ_{ref} which is the absolute refractory period (Fig. 1).

The total synaptic current is given by the sum of glutamatergic excitatory components (AMPA and NMDA) and inhibitory components (GABA). As we described above, we consider that external excitatory contributions are produced through AMPA receptors ($I_{\mathrm{AMPA,ext}}$), while the excitatory recurrent synaptic currents are produced through AMPA and NMDA receptors ($I_{\mathrm{AMPA,rec}}$ and $I_{\mathrm{NMDA,rec}}$). The total synaptic current is therefore given by:

$$I_{\mathrm{syn}}(t) = I_{\mathrm{AMPA,ext}}(t) + I_{\mathrm{AMPA,rec}}(t) + I_{\mathrm{NMDA,rec}}(t) + I_{\mathrm{GABA}}(t) \tag{2}$$

where the current generated by each receptor type follows the general form:

$$I(t) = g(V(t) - V_E) \sum_{j=1}^{N} w_j s_j(t) \tag{3}$$

and $V_E = 0$ mV for the excitatory (AMPA and NMDA) synapses and -70 mV for the inhibitory (GABA) synapses. The synaptic strengths w_j are specified by the architecture. The time course of the current flow through each synapse is dynamically updated to describe its decay by altering the fractions of open channels s according to equations with the general form:

$$\frac{ds_j(t)}{dt} = -\frac{s_j(t)}{\tau} + \sum_k \delta(t - t_j^k) \tag{4}$$

where the sum over k represent a sum over spikes emitted by presynaptic neuron j at time t_j^k, and τ is set to the time constant for the relevant receptor. In the case of the NMDA receptor, the rise time as well as the decay time is dynamically modeled, as it is slower. Details are provided by [10].

The problem now is how to analyze the dynamics and how to set the parameters which are not biologically constrained by experimentally determined values. The standard trick is to simplify the dynamics via the *mean-field approach* at least for the stationary conditions, i.e., for periods after the dynamical transients, and to analyze there exhaustively the bifurcation diagrams of the dynamics. A bifurcation diagram shows the possible dynamical states of the system as a function of the model-parameters. This enables a posteriori selection of the parameter region which shows in the bifurcation diagram the emergent behavior that we are looking for (e.g., sustained delay activity, biased competition, etc.). After that, with this set of parameters, we perform the full non-stationary simulations using the *true dynamics* only described by the full integrate-and-fire scheme. The mean-field approximation represents a well-established way of exploring the behavior of the network [33, 14, 18, 17, 3, 37, 36]. It assures that the dynamics of the network will converge to a stationary attractor which is consistent with the asymptotic behavior of an asynchronous firing network of IF neurons [5, 19, 16].

In the standard mean-field approach, the network is partitioned into populations of neurons which share the same statistical properties of the afferent currents (currents that enter each neuron), and fire spikes independently at the same rate. The essence of the mean-field approximation is to simplify the integrate-and-fire equations by replacing, in accordance with the diffusion approximation [35], the sums of the synaptic components by the average DC component and a fluctuation term. The stationary dynamics of each population can be described by the *population transfer function* $F()$, which provides the average population rate as a function of the average input current. The set of stationary, self-reproducing rates ν_i for the different populations i in the network can be found by solving a set of coupled self-consistency equations:

$$\nu_i = F(\mu_i(\nu_1, ..., \nu_N), \sigma_i(\nu_1, ..., \nu_N)) \tag{5}$$

where $\mu_i()$ and $\sigma_i()$ are the mean and standard deviation of the corresponding input current, respectively. To solve these equations, a set of first-order differential equations, describing a *fake dynamics* (in contrast to the 'true' underlying spiking dynamics) of the system, whose fixed point solutions correspond to the solutions of Eq. 5, is used :

$$\tau_i \frac{d\nu_i(t)}{dt} = -\nu_i(t) + F(\mu_i(\nu_1, ..., \nu_N), \sigma_i(\nu_1, ..., \nu_N)) \quad (6)$$

The standard mean-field approach neglects the temporal properties of the synapses, i.e., considers only delta-like spiking input currents. Consequently, after this simplification, the transfer function $F()$ is an Ornstein-Uhlenbeck solution for the simplified integrate-and-fire equation $\tau_x \frac{dV(t)}{dt} = -V(t) + \mu_x + \sigma_x \sqrt{\tau_x}\eta(t)$, as detailed by [5]. An extended mean-field framework which is consistent with the integrate-and-fire and synaptic equations described above, i.e., that considers both the fast and slow glutamatergic excitatory synaptic dynamics (AMPA and NMDA) and the dynamics of GABA-inhibitory synapses, were derived by [5]. The mean-field analysis performed in this work uses the formulation derived in [5, 4], which is consistent with the network of neurons used.

To model a cortical area, a network of interconnected excitatory and inhibitory neurons is considered. The basic general network architecture used implements a multi-modular system where stimuli-related inputs are processed in the context of neuronal reverberation, cooperation and competition biased by task-relevant information.

The structure of the network consists of a set of distinct pools of neurons which are defined by shared inputs and weights of the connections. We consider three general types of pools: pools of excitatory neurons which selectively encode information associated with some given task to be modeled (selective pools); one pool of excitatory neurons which are not directly activated in association with the task at hand (non-selective pool) and one pool of inhibitory neurons. All neurons receive a background external input assumed to originate in neurons not explicitly modeled. Added to this noise input, the selective pools can receive external inputs which are task-related. These inputs convey into the model either the presentation of stimuli or an attentional state or context knowledge. This last type of inputs have the ability to bias the processing taking place in the network and are assumed to be encoded in cortical areas not explicitly modeled.

The framework of biased competition and cooperation is implemented by differentially setting the strengths of the interneuronal synaptic connections. These connection strengths or weights describe relative departures of the synaptic conductivities from their average value over the network. The weights are considered fixed, after some tuning process (possibly a Hebbian rule) which is not simulated. It is reasonable to assume that neurons within the same selective pool are strongly coactivated and hence that, following a Hebbian learning rule, the synaptic strength of the connections between such neurons (pool cohesion) are higher than average. These strong weights implement reverberation of the intra-pool neuronal activities, which can underly formation of working memory, that is appearance of sustained high activity coding for a previously presented stimuli, which is absent at present. Some selective pools are involved in processing related information (for example, similar inputs) in the context of a certain task. Neurons belonging to these pools are likely to have correlated activity and therefore are assumed to be linked through connections with stronger-than-average weights, implementing cooperation between these pools. The selective pools which are activated by different stimuli or tasks are likely to

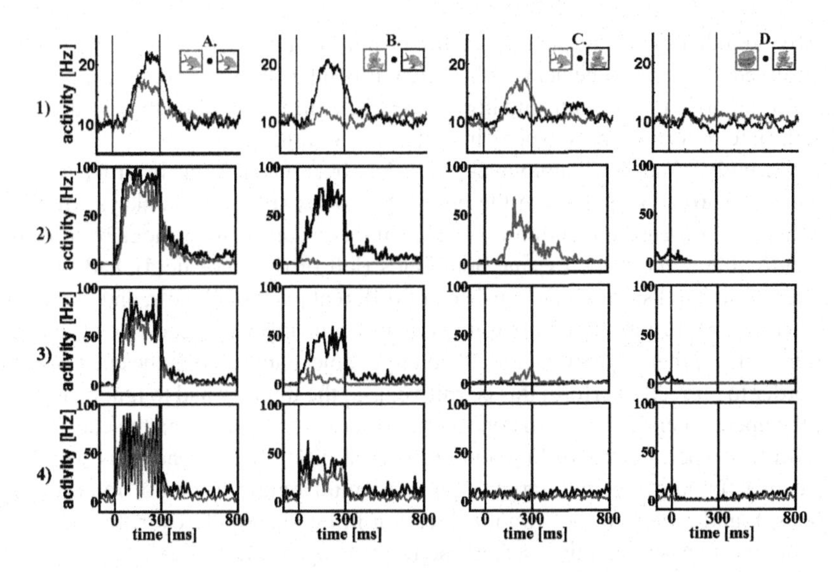

Fig. 2. 1st row: Attentional filtering experiment. Each pair of traces shows the mean neuronal responses to a bilateral stimulus shown in the inset. (A) target/target, (B) Non-target/target, (C) target/ non-target, and (D) non-target/non-target stimulus pairs. Black and gray trace are recorded, when the attention was drawn to the preferred location (black box around stimulus) or to the non-preferred location. For each pair of traces, the stimulus was identical, but the response reflected only the identity of the object in the attended location. 2nd row: Simulation result of a spiking network with biased competition and cooperation. 3rd row: Simulation result for a weak increase of dopamine D2 receptor activity. 4th row: Result for strong increase of D2 activity.

have uncorrelated activities and thus be connected through synaptic weights weaker than average. These selective pools influence each other's activities mainly through the inhibitory neurons, which then implement a mechanism of global competition between such excitatory pools.

The theoretical framework for biased competition and cooperation networks described here has been used to model single neuronal responses fMRI activation patterns, psychophysical measurements, effects of pharmacological agents and of local cortical lesions [29, 9, 10, 34]. The examples described in the next three sections focus mainly on results of neurophysiological experiments.

3 Attentional filtering

Mechanisms of selective attention form an important basis of cognitive processes. By attention, information is selected and filtered out in a context-dependent way. The context is provided by the internal state of the brain, which can represent current hypotheses about the environment. For example, visual information can be selectively attended to depending on whether or not this input is relevant for a task to be subserved or a goal to be achieved. Hence, it is interesting to investigate how attention could arise in the brain, and how it can be flexibly controlled by the surrounding brain state reflecting the subject's hypothesis about the world.

Recently, a neurophysiological study has been carried out in order to investigate the neuronal responses in the prefrontal cortex (PFC) of monkeys which were engaged in a focused attention task [15]. In this experiment, a monkey, after being cued to attend one of two visual hemifields (left or right eye-field), had to watch a series of bilateral stimuli that consisted of different pairs of objects, and to react with a saccade (rapid intermittent eye movement occurring when eyes fix on one point after another) if and only if a predefined target object appeared in the cued hemifield. In order to correctly perform this cognitive task, the monkey had to ignore any object in the uncued hemifield and to concentrate (focus his attention) on the cued location.

In the experiment, Everling and co-workers first observed neurons which were selective for target or non-target stimuli (stimuli requiring or not requiring a response), and preferred the stimulus location in one hemifield over the other. Next, during the focused attention task, these neurons again discriminated between target and non-target, but only when the stimulus changed in the attended location (Fig. 2, row 1). The stimulus in the non-attended location had no influence on the neuronal response. This effect we refer to as attentional filtering. In other words, the context, which is given by the relevance of a stimulus for the task, acts in a multiplicative way on its representation in the PFC. Only a task-relevant stimulus (i.e., target in the cued hemifield) is gated by the context and is allowed to be represented.

Attentional filtering represents a particularly strong attentional effect, in which the context gates sensory input in an all-or-none fashion. Moreover, attentional filtering might be part of a neural correlate of inattentional blindness, which is the inability of humans to recover any information from unattended sensory stimuli [31]. Inattentional blindness is thought to be part of an important cognitive mechanism, namely that of focusing or 'concentrating' on a task to be performed.

Motivated by these observations, we built a neurodynamical model to investigate, how this strong attentional effect can arise from a weak modulatory bias which mediates the cortical context [34]. The model is set up as a layer of spiking neurons which are grouped into four selective pools, one non-selective pool and one inhibitory pool, as described in the previous section. Neurons of each selective pool not only share the same weight, but in addition receive the same bottom-up sensory input. The input selectively encodes whether there was a target or non-target in the left or right hemifield. Correspondingly, the four selective pools are denoted according to their inputs as target-left (TL), target-right (TR), other left (OL) and other right (OR). In agreement with the Hebbian learning rule, weights within each pool were strongest. Weights between the two target and the two non-target pools are intermediate, because these pairs receive similar inputs. These weights are found to mediate cooperation between same-object pools. Weights between all other neurons are weaker than average, because they receive different inputs. The latter weights are found to implement competition between pools for activity. The selective pools receive two top-down inputs which act as bias. The first top-down signal biases neurons that are selective for the target object. The second top-down signal facilitates neurons which have the cued location as preferred location. Both biases together can provide sufficient information about the task to be solved. They are found to guide the competition and cooperation process.

The second row of Fig. 2 shows the pool-averaged responses of the model TR neurons for the same stimulus conditions and attentional states as the experimental results of row 1. It can be seen that the model traces are in good agreement with the experimental results. This demonstrates that a weak attentional bias can be strongly and selectively amplified by cooperation and competition and lead to an all-or-none attentional filtering effect. Simulating with different parameters, it can be shown that in the framework of the present model both cooperation and competition are needed together in order to reproduce the effect.

Departing from the parameter setting which correctly reproduces the experiment, one can also formulate quantitative predictions of the effect of neurotransmitters or pharmacological treatment. This will be exemplified by modeling the effect of an increase in dopamine concentration on attentional filtering. An increase in dopamine concentration, accompanied by an increase in D2 receptor activation, is known to decrease both NMDA and GABA conductances [21, 38]. Hence, a weak increase in D2 receptor activation was modeled by multiplying both NMDA and GABA conductances by a factor of 0.7, a strong increase by multiplying the same quantities with 0.2. The model neuronal responses are shown in rows 3 and 4 of Fig. 2, respectively. We found that when the dopamine level increases slightly, the response to a weak stimulus is no longer attended to (Fig. 2, 3C). This might be related to a degraded ability to shift selective attention to a new, non-prominent stimulus. As the level of dopamine is further increased, attentional effects become more and more impaired in general (Fig. 2, row 4). Hence, by using a neurodynamical model, we can predict that an increase in dopamine concentration will lead to a progressive weakening of the attentional filtering effect.

4 Working memory

Many real world behaviors depend on arbitrary stimulus-response mappings. To explore the role of the prefrontal cortex in stimulus-response associations, the experiment of Assad, Rainer and Miller [1] is designed to combine aspects of cue-response learning with an object-response task. We suggest a neurodynamical model based on the finding of object-direction selective neurons in the experiment, which allows us to draw further conclusions about the possible underlying neural substrate.

The delayed-response experiment starts with presentation of an object A or B, and after a delay a response is given by a leftward or rightward saccadic eye movement according to the object identity. To this simple object-response task an associative aspect is added by incorporating a reward learning-mechanism with two possible conditions: One associates object A with a leftward motor response L, B with a rightward response R (direct rule), and vice versa (reversal rule). During the experi-

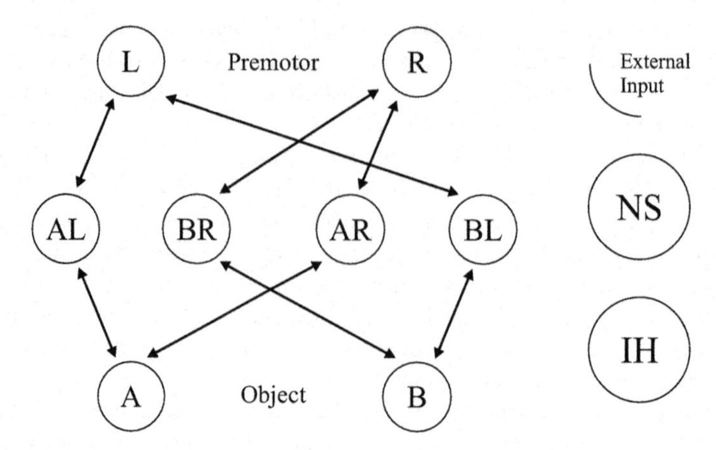

Fig. 3. The basic setup of the network model consists of 8 selective pools (A,B,AL,BR,AR,BL,L,R), a non-selective pool (NS) and an inhibitory pool (IH). The external environment of the neuronal assembly is modeled with an external input.

ment the current condition is indicated by reward learning, so that the monkey knows which rule applies.

During the trials, neurons in the lateral prefrontal cortex of the monkeys are recorded to assess the underlying neuronal basis of the conditional visuomotor tasks. The interesting finding is that, besides neurons that selectively respond to either the cue or the associated response, there are object-response selective neurons that respond to cue and response properties in a nonlinear fashion: High activity is displayed during object A with response L, while there are low activity levels in all other combinations.

Based on the single-cell findings of [1] we postulate 8 different kinds of pools of neurons: Pools that are selective to the presented objects, designated by A and B; ones that code the output in terms of the rightward and leftward motor response, called L and R, respectively; inspired by the nonlinear object-response selective neurons found, we also introduce all possible combinations of object and premotor selectivity, denoted: AL, AR, BL and BR. The neuronal basis of the model is made up of these 8 selective pools, a non-selective one and an inhibitory pool for global inhibition.

The connections between the pools are assumed to have been built up according to the Hebbian learning rule, which leads to a hierarchical neurodynamical model: There are three layers, an input layer with the object pools A,B, a premotor layer with the premotor pools L,R and an intermediate layer in between that contains pools for all possible combinations. We associate A with AL and AR, also B with BL and BR. The same is done for the premotor neurons: L with AL, BL and R with AR, BR (Fig. 3).

The learned associations are modeled with a higher external input to intermediate neurons, which mediate most likely the mapping between the object-selective and premotor neurons. In terms of neurodynamics, this external influence is a bias that shifts the dynamics of the system in order to account for a learned rule.

The analysis is conducted by means of a mean-field exploration of the parameter space that is spanned by the free parameters in the network. These parameters are the pool cohesion, the external bias for the applied rule and the structural feed forward and feed backward connections within the network. The goal of the analysis is to find a non-linear mapping between object and response in accordance with the applied rule (direct or reversal) that has also been identified in the experiment of [1].

In the analysis we find that the pool cohesion and the strength of the external bias are the important factors in order to show biased competition properties throughout the network, while the structural parameters, the feed-forward and feed-backward parameters influence the way in which biased competition takes effect.

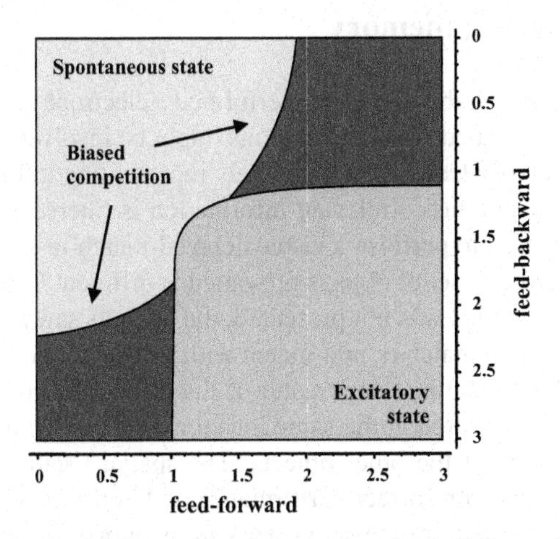

Fig. 4. Qualitative behavior of the system when varying the feed-forward and feed-backward connection strength while leaving the rest of the parameters (pool-cohesion, external input and external bias) at fixed values. Three basic states are found: A spontaneous state in which the presentation of the stimuli to the network does not show any effect, an excitatory state which presents overall high activity and a biased competition state in which the presentation of the object and the external bias interact.

Fig. 4 shows conceptual regions found by varying the relative feed-forward and -backward connection strengths. With low relative connection strengths we observe that the presentation of the object stimulus does not have any effect and the pools just show spontaneous activity, although we present a stimulus to the network. On the contrary, strong connection strengths lead to a high excitation of all pools, although the implemented bias that should account for the learned rule (direct/reversal) does not effect the final state of the neurodynamics. In an area with moderate connection strength which is also characterized by either stronger feed-forward or feed-backward connections, we observe various distinct kinds of behavior depending on the specific connectivity properties which we summarized in the biased competition region in Fig. 4. A thorough analysis of the surrounding parameter space reveals multiple dependencies in the network that need to be met in order to observe the desired qualitative behavior.

In summary, we built up and analyzed a neurodynamical model based on the experimental data of [1]. We have been able to reproduce the experimental results by postulating corresponding pools of neurons and using the biased competition and cooperation hypothesis. Furthermore, our results suggest that the network has to be in a finely tuned dynamical equilibrium, otherwise it is not able to show the desired results. The figure also suggests that the connection strengths have to be stronger in one direction, either feed-forward or backward, in order to match the measured biological data of the underlying experiment.

5 Selective working memory

Recently, Rainer and collaborators [26] performed a electrophysiological experiment in monkeys which revealed that PFC neurons might be involved not only in the storage of working memory but also in selectively representing information relevant for task performance, while task-irrelevant information is filtered out. In this study the monkeys were required to perform a visual delayed-match-to-sample task. The sample stimuli consisted of a set of objects presented at different locations. After a delay period, where the stimulus was not presented, the animals saw a new stimulus which they had to identify as a match or non-match with respect to the previously presented sample. A stimuli was defined as a match if the target object (whose identity was known beforehand) appeared at the same location as in the sample period, irrespectively of the locations of the other objects. The spatial locations of the non-target objects were irrelevant for correct performance and hence it was not necessary for the monkey to store this information in working memory. The experimental results showed that neurons in PFC exhibited delay activity which coded selectively for task relevant information (position or location of the target object). The task irrelevant features of the sample stimulus did not influence the measured delay activity. More, although the task was defined through object identity, it not only determined which object identity information should be retained in working memory but also which spatial information to store. These experimental findings can be explained within the context of biased competition and cooperation, using a neurodynamical computational model as described above. However, to fully account for the experimental results, an extension of the general model had to be introduced, allowing the implementation of a mechanism we called *modular* biased competition.

The specific network architecture used is shown in Fig. 5. For simplicity we considered only stimuli consisting of two possible objects presented in two possible positions. Four neuronal pools (OS) receiving stimulus specific external inputs were considered. Neurons in these pools were assumed to code for a particular object and position. Apart from this structure, we considered pools encoding just for object (O) or spatial location (S), since the experimental results reported were measured from neurons with such responsiveness. In each trial, one of the O pools received exter-

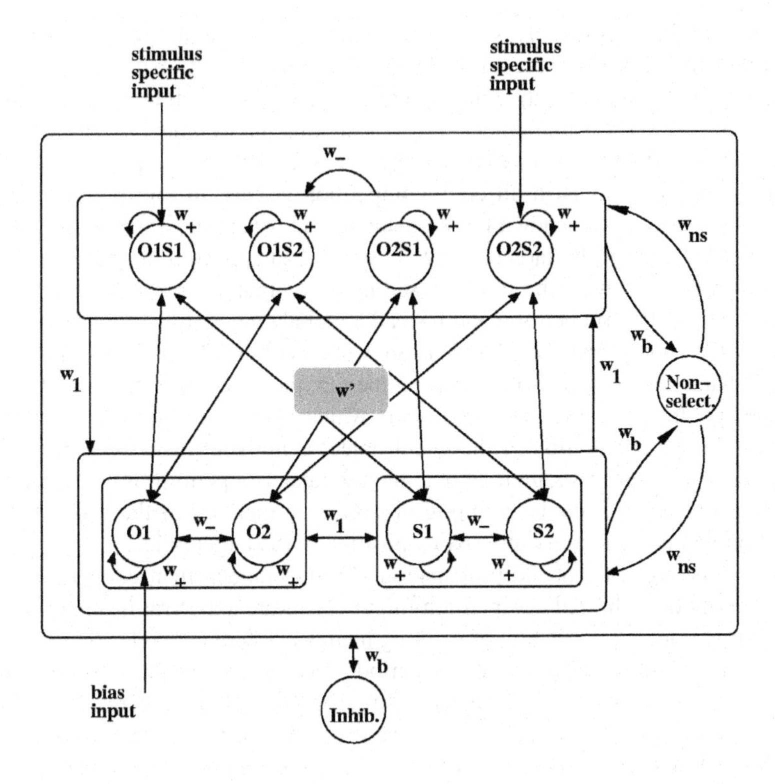

Fig. 5. Network architecture.

nal input coding for identity of the target object, which biased the processing by the network of the stimulus specific inputs. With the rational explained above, the weight (here $w_+ = 2.1$) (see Fig. 5) between neurons of the same selective pool was considered to be the highest weight, implementing formation of working memory. Cooperation between pools coding for the same stimuli was implemented by taking the corresponding weights (here $w' = 1.8$) larger than the average weight, while a value of w_- (here 0.3) smaller than 1 implemented competition. The mechanism of biased competition implemented the filtering out of irrelevant features for task performance that is the selection of which information to be stored, according to context. The cooperation was shown to be essential for the ability to select what spatial information to maintain, based on information about object identity. However, to account for the experimental results a modification of the biased competition and cooperation model was important. This extension implemented intracortical modular competition between three separate groups of selective pools of neurons (O pools, S pools and OS pools), as opposed to a global competition between all selective pools. This mechanism together with cooperation allowed the propagation of competition through the different modules of the network and the subsequent formation of selective spatial working memory. This was done by letting one spatial selective pool win a local competition, instead of resulting in global winners. The global winners would be among the O and OS selective pools, which receive a larger total external input. Modular competition was implemented by considering the weights (w_{-}) of the connections between pools which should compete weaker than the weights (w_1) of the connections between pools in different competition groups. In Fig. 6 results from mean-field simulations are shown for the delay period, where the monkey should remember the location of the target (object 1) which had appeared in space 1 during stimulus presentation. It can be seen that setting the weight w_1 lower than w_{-} (0.3), destroys the modular competition and instead a global competition is implemented. In this case the pools O1S1, O2S2 and O1 are winners of the global competition (see Fig. 6 left-hand side, both the plots and, on the top, the schematic representation of the winner pools). There are no space pools which exhibit sustained high activity and hence the spatial location of the target object is not kept in working memory. For values of w_1 at least as high as w_{-}, the modular competition is implemented and propagated to the S pools, resulting in storage in memory of the task relevant information that is the identity and spatial position of the target object.

Our results suggest that the propagation, through intracortical cooperation, of intracortical modular biased competition might constitute a general mechanism for implementing selective formation of working memory, where task-relevant information is maintained in mind over a delay period while task-irrelevant information is filtered out through an attentional mechanism. The network architecture described in this section is an extension of the general network structure used to implement biased competition and cooperation, which was used in the examples above.

The extension introduced implements competition among subsets of neuronal pools as opposed to global competition. This mechanism was shown to be important for the formation of selective working memory at the level of one feature dimension (space) although the relevant information for task performance was defined at the level of another feature dimension (object identity).

6 Future directions

We have to keep in mind that the models include several abstractions and simplifications that are on the one hand necessary due to computational constraints and on the other hand an essential part of the modeling process. If one wants to find out which network components are needed to show specific properties they have to be studied separately. Therefore the overall goal should not be to build up a highly comprehensive model of the brain, but to try to extract neurodynamical effects that are responsible for single functions of the brain. The second step would be to put them together to achieve a holistic picture of brain functionality. The presented models should not be regarded as given and fixed but as an open framework which allows to extract important information about the neurodynamical effects underlying higher cognitive brain functions.

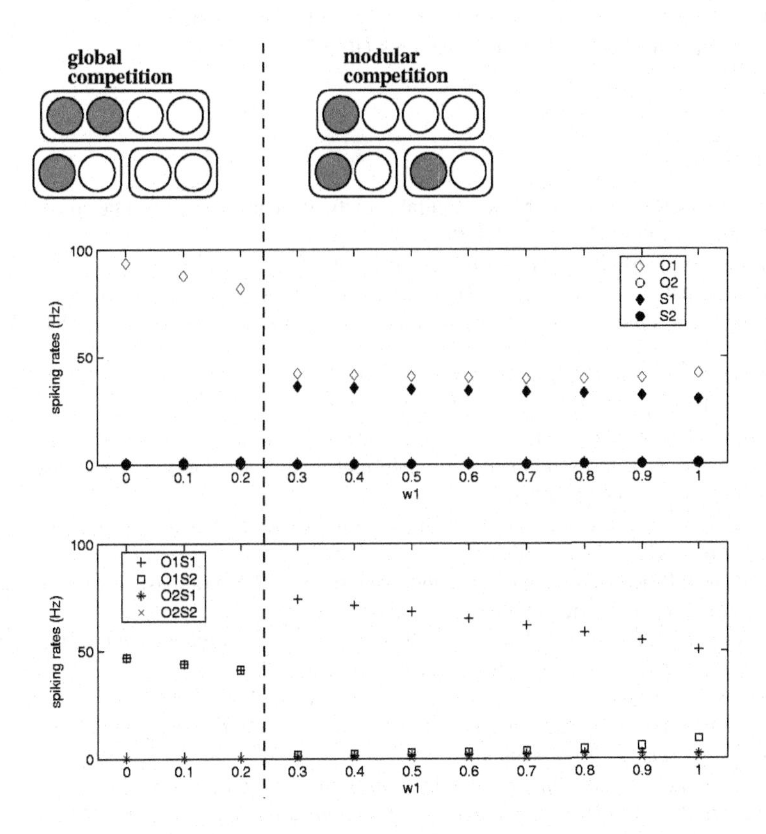

Fig. 6. Spiking rates of the selective pools, as a function of w_1. On the top schematic representation of the neuronal pools showing high sustained activity during the delay period, hence coding for information stored in working memory.

There are several parts of the models which could be altered and discussed in the process of discussing certain effects: Full vs. partial connectivity, local vs. global inhibition, influence of the external neurons, modeling and selection of neurotransmitters.

Even though many neural and synaptic mechanisms are not known exactly, we can make explicit assumptions for these mechanisms, analyze them and thereby draw specific conclusions about the underlying neural behavior.

Acknowledgments

This work was supported by the German Ministry of Research, BMBF grant 01IBC01A and through the European Union, grant IST-2001-38099. M. Loh was supported by the International Graduate School of Catalonia, R. Almeida was financed by a

Marie Curie Individual Fellowship, QLK6-CT-2002-51439 and G. Deco by Institucion Catalana de Recerca i Estudis Avançats (ICREA).

References

1. W. Assad, G. Rainer, and E. Miller. Neural activity in the primate prefrontal cortex during associative learning. *Neuron*, 21:1399–1407, 1998.
2. A. Baddeley. *Working Memory*. Oxford University Press, London, 1986.
3. R. Ben-Yishai, R. Lev Bar-Or, and H. Sompolinsky. Theory of orientation tuning in visual cortex. *Proceedings of the National Academy of Sciences of the USA*, 92(9):3844–3848, 1995.
4. N. Brunel and S. Sergi. Firing frequency of leaky integrate-and-fire neurons with synaptic currents dynamics. *J. Theor. Biol.*, 195:87–95, 1998.
5. N. Brunel and X. Wang. Effects of neuromodulation in a cortical networks model of object working memory dominated by recurrent inhibition. *Journal of Computational Neuroscience*, 11:63–85, 2001.
6. L. Chelazzi. Serial attention mechanisms in visual search: a critical look at the evidence. *Psychol Res.*, 62:195–219, 1999.
7. L. Chelazzi, E. K. Miller, J. Duncan, and R. Desimone. A neural basis for visual search in inferior temporal cortex. *Nature*, 363:345–347, 1993.
8. P. Dayan and L.F. Abbott. *Theoretical Neuroscience: Compuational and Mathematical Modeling of Neural Systems*. MIT Press, Cambridge, 2001.
9. G. Deco, E. T. Rolls, and B. Horwitz. "what" and "where" in visual working memory: A computational neurodynamical perspective for integrating fmri and single-neuron data. *J. Cogn. Neurosci.*, 2003. in press.
10. G. Deco and E.T. Rolls. *Attention and Working Memory: A Dynamical Model of Neuronal Activity in the Prefrontal Cortex, accepted for publication, in press*. European Journal of Neuroscience, Oxford, 2003.
11. R. Desimone and J. Duncan. Neural mechanisms of selective visual attention. *Annual Review of Neuroscience*, 18:193–222, 1995.
12. J. Duncan. Cooperating brain systems in selective perception and action. In T. Inui and J. L. McClelland, editors, *Attention and Performance XVI*, pages 433–458. MIT Press, Cambridge MA, 1996.
13. J. Duncan and G. Humphreys. Visual search and stimulus similarity. *Psychol Rev.*, 96:433–58, 1989.
14. J. Eggert and J. L. van Hemmen. Unifying framework for neuronal assembly dynamics. *Physical review E*, 61(2):1855–1874, 2000.
15. S. Everling, C. Tinsley, D. Gaffan, and J. Duncan. Filtering of neural signals by focused attention in the monkey prefrontal cortex. *Nature Neurosci.*, 5:671–676, 2002.
16. S. Fusi and M. Mattia. Collective behavior of networks with linear (vlsi) integrate and fire neurons. *Neural Comput.*, 11:633–652, 1999.
17. W. Gerstner. Time structure of the activity in neural network models. *Physical review E*, 51:738–758, 1995.
18. W. Gerstner. Population dynamics of spiking neurons: fast transients, asynchronous states and locking. *Neural Computation*, 12(1):43–89, 2000.
19. P. Del Giudice, S. Fusi, and M. Mattia. Modeling the formation of working memory with networks of integrate-and-fire neurons connected by plastic synapses. *J. Physiol. Paris*, 2003. in press.

20. E. R. Kandel and J. Schwartz, editors. *Principles of neural science.* Elsevier, second edition, 1985.
21. D. Law-Tho, J. C. Hirsch, and F. Crepel. Dopamine modulation of synaptic transmission in rat prefrontal cortex: an in vitro electrophysiological study. *Neurosci Res.*, 21:151–60, 1994.
22. N. K. Logothetis, J. Pauls, M. Augath, T. Trinath, and A. Oeltermann. Neurophysiological investigation of the basis of the fmri signal. *Nature*, 412:150–7, 2001.
23. E. K. Miller, P. M. Gochin, and C. G. Gross. Suppression of visual responses of neurons in inferior temporal cortex of the awake macaque by addition of a second stimulus. *Brain Res.*, 616:25–9, 1993.
24. J. Moran and R. Desimone. Selective attention gates visual processing in the extrastriate cortex. *Science*, 229:782–784, 1985.
25. B. C. Motter. Focal attention produces spatially selective processing in visual cortical areas v1, v2, and v4 in the presence of competing stimuli. *J Neurophysiol.*, 70:909–19, 1993.
26. G. Rainer, W. F. Asaad, and E. K. Miller. Selective representation of relevant information by neurons in the primate prefrontal corte. *Nature*, 393(6685):577–579, 1998.
27. J. Reynolds and R. Desimone. The role of neural mechanisms of attention in solving the binding problem. *Neuron*, 24:19–29, 1999.
28. E. T. Rolls and A. Treves. *Neural networks and brain function.* Oxford University Press, 1998.
29. E.T. Rolls and G. Deco. *Computational Neuroscience of Vision.* Oxford University Press, Oxford, 2002.
30. G. M. Sheperd, editor. *The synaptic organization of the brain.* Oxford University Press, fourth edition, 1998.
31. D. J. Simons. Attentional capture and inattentional blindness. *Trends Cognit. Sci.*, 4:147–155, 2000.
32. R. Spitzer, H. Desimone and J. Moran. Increased attention enhances both behavioral and neuronal performance. *Science*, 240:338–340, 1988.
33. M. Stetter. *Exploration of Cortical Function.* Kluwer Scientific Publishers, Dordrecht, 2002.
34. M. Szabo, R. Almeida, G. Deco, and M. Stetter. Cooperation and biased competition model can explain attentional filtering in the prefrontal cortex. *Eur. J. Neurosci*, 19(7):1969–1977, 2004.
35. H. Tuckwell. *Introduction to theoretical neurobiology.* Cambridge University Press, Cambridge, 1988.
36. H. R. Wilson and J. D. Cowan. Excitatory and inhibitory interactions in localized populations of model neurons. *Biophysical Journal*, 12(1):1–24, 1972.
37. H. R. Wilson and J. D. Cowan. A mathematical theory of the functional dynamics of cortical and thalamic nervous tissue. *Kybernetik*, 13(2):55–80, 1973.
38. P. Zheng, X. X. Zhang, B. S. Bunney, and W. X. Shi. Opposite modulation of cortical n-methyl-d-aspartate receptor-mediated responses by low and high concentrations of dopamine. *Neuroscience*, 91:527–35, 1999.

Index